Advance Numerical Techniques to Solve Linear and Nonlinear Differential Equations

RIVER PUBLISHERS SERIES IN MATHEMATICAL, STATISTICAL AND COMPUTATIONAL MODELLING FOR ENGINEERING

Series Editors:

Mangey Ram
Graphic Era University, India

Tadashi Dohi
Hiroshima University, Japan

Aliakbar Montazer Haghighi
Prairie View Texas A&M University, USA

Applied mathematical techniques along with statistical and computational data analysis has become vital skills across the physical sciences. The purpose of this book series is to present novel applications of numerical and computational modelling and data analysis across the applied sciences. We encourage applied mathematicians, statisticians, data scientists and computing engineers working in a comprehensive range of research fields to showcase different techniques and skills, such as differential equations, finite element method, algorithms, discrete mathematics, numerical simulation, machine learning, probability and statistics, fuzzy theory, etc.

Books published in the series include professional research monographs, edited volumes, conference proceedings, handbooks and textbooks, which provide new insights for researchers, specialists in industry, and graduate students.

Topics included in this series are as follows:-

- Discrete mathematics and computation
- Fault diagnosis and fault tolerance
- Finite element method (FEM) modeling/simulation
- Fuzzy and possibility theory
- Fuzzy logic and neuro-fuzzy systems for relevant engineering applications
- Game Theory
- Mathematical concepts and applications
- Modelling in engineering applications
- Numerical simulations
- Optimization and algorithms
- Queueing systems
- Resilience
- Stochastic modelling and statistical inference
- Stochastic Processes
- Structural Mechanics
- Theoretical and applied mechanics

For a list of other books in this series, visit www.riverpublishers.com

Advance Numerical Techniques to Solve Linear and Nonlinear Differential Equations

Editors

Geeta Arora

Lovely Professional University, Punjab, India

Mangey Ram

Graphic Era (Deemed to be University), Dehradun, India

NEW YORK AND LONDON

Published 2024 by River Publishers
River Publishers
Alsbjergvej 10, 9260 Gistrup, Denmark
www.riverpublishers.com

Distributed exclusively by Routledge
605 Third Avenue, New York, NY 10017, USA
4 Park Square, Milton Park, Abingdon, Oxon OX14 4RN

Advance Numerical Techniques to Solve Linear and Nonlinear Differential Equations / by Geeta Arora, Mangey Ram.

© 2024 River Publishers. All rights reserved. No part of this publication may be reproduced, stored in a retrieval systems, or transmitted in any form or by any means, mechanical, photocopying, recording or otherwise, without prior written permission of the publishers.

Routledge is an imprint of the Taylor & Francis Group, an informa business

ISBN 978-87-7022-987-6 (hardback)
ISBN 978-87-7004-057-0 (paperback)
ISBN 978-10-0381-100-8 (online)
ISBN 978-10-3263-029-8 (master ebook)

While every effort is made to provide dependable information, the publisher, authors, and editors cannot be held responsible for any errors or omissions.

Contents

Preface	ix
List of Figures	xi
List of Tables	xiii
List of Contributors	xv
List of Abbreviations	xvii

1 A Slow Varying Envelope of the Electric Field is Influenced by Integrability Conditions   1
Mostafa M. A. Khater
- 1.1 Introduction 2
- 1.2 Solitary Wave Solutions 6
  - 1.2.1 The Khater II method's results 6
  - 1.2.2 The Sardar sub-equation method's results 8
- 1.3 Results and Discussion 15
- 1.4 Conclusion 15

2 Novel Cubic B-spline Based DQM for Studying Convection–Diffusion Type Equations in Extended Temporal Domains   21
J. P. Shukla, B. K. Singh, C. Cattani, and M. Gupta
- 2.1 Introduction 22
- 2.2 Portrayal of nHCB-DQM 23
- 2.3 Computation of Wt. Coeff. $a_{il}^{(1)}$ and $a_{il}^{(2)}$ 25
- 2.4 The nHCB-DQM for the Class of C–D Eqn 26
- 2.5 Numerical Results and Discussion 27
- 2.6 Conclusion 31

3 Study of the Ranking-function-based Fuzzy Linear Fractional Programming Problem: Numerical Approaches — 37
Ravinder Kaur, Rakesh Kumar, and Hatıra Günerhan
- 3.1 Introduction — 37
- 3.2 Preliminaries — 38
- 3.3 General Form of Fuzzy LFPP — 40
- 3.4 Algorithm for the Solution of FLFPP with Trapezoidal Fuzzy Number TrpFN — 41
- 3.5 Numerical Example — 41
- 3.6 Conclusion — 43

4 Orthogonal Collocation Approach for Solving Astrophysics Equations using Bessel Polynomials — 45
Shelly Arora and Indu Bala
- 4.1 Introduction — 45
- 4.2 Bessel Collocation Method — 47
- 4.3 Convergence Analysis — 55
- 4.4 Numerical Examples — 56
- 4.5 Conclusions — 59

5 B-spline Basis Function and its Various Forms Explained Concisely — 63
Shubham Mishra, Geeta Arora, and Homan Emadifar
- 5.1 Introduction — 64
  - 5.1.1 Idea of spline — 64
- 5.2 B-spline — 64
  - 5.2.1 Trigonometric B-spline — 68
    - 5.2.1.1 Three degree or cubic trigonometric B-spline — 68
  - 5.2.2 Hyperbolic B-spline — 69
    - 5.2.2.1 Cubic hyperbolic B-spline — 69
  - 5.2.3 Uniform algebraic trigonometric tension B-spline — 70
  - 5.2.4 Exponential B-spline — 72
    - 5.2.4.1 Exponential cubic B-spline — 73
  - 5.2.5 Quartic hyperbolic trigonometric B-spline — 74
  - 5.2.6 Quintic hyperbolic B-spline — 74
  - 5.2.7 Modified cubic UAH (uniform algebraic hyperbolic) tension B-spline — 76
  - 5.2.8 Modified cubic UAT tension B-spline — 77

		5.2.9 Quintic trigonometric B-spline	78
		5.2.10 Quartic trigonometric differential	79
	5.3	Equation Solved by the B-spline Basis Function	79
	5.4	Conclusion	80

6 A Comparative Study: Modified Cubic B-spline-based DQM and Sixth-order CFDS for the Klein−Gordon Equation   85
G. Arora, B. K. Singh, Neetu Singh, and M. Gupta

6.1	Introduction		86
6.2	Methodology		87
	6.2.1	MCB-DQM	88
		6.2.1.1 The weighting coefficients	88
	6.2.2	CFDS6	89
6.3	Implementation of the Method		91
6.4	Results and Discussion		91
6.5	Conclusion		97

7 Sumudu ADM on Time-fractional 2D Coupled Burgers' Equation: An Analytical Aspect   103
Mamta Kapoor

7.1	Introduction	103
7.2	Main Text Implementation of the Scheme	104
7.3	Examples and Calculation	106
7.4	Graphs and Discussion	113
7.5	Concluding Remarks	114

8 Physical and Dynamical Characterizations of the Wave's Propagation in Plasma Physics and Crystal Lattice Theory   117
Raghda A. M. Attia and Mostafa M. A. Khater

8.1	Introduction		118
8.2	GP model's traveling wave solutions		122
	8.2.1	Solitary wave solutions	122
	8.2.2	Solution's accuracy	124
8.3	Soliton Solution's Novelty		124
8.4	Conclusion		129

9 Numerical Solution of Fractional-order One-dimensional Differential Equations by using a Laplace Transform with the Residual Power Series Method   135

Rajendra Pant, Geeta Arora, Manik Rakhra, and Masoumeh Khademi
- 9.1 Introduction 136
- 9.2 Preliminaries 137
- 9.3 Methodology 139
- 9.4 Numerical Solutions 141
- 9.5 Conclusion 145

Index   149

About the Editors   151

Preface

Mathematics plays a vital role in the resolution of numerous physical and material world problems. Differential equations have emerged in recent years as a result of mathematical modeling in several fields of engineering and biology. Significant advancements have been made in the development of computational approaches by mathematicians. Real-world applications of contemporary science and engineering necessitate the use of computational approaches into the study of differential equations.

Numerical analysis is the discipline of mathematics dealing with the theoretical foundations of numerical algorithms for the solving of scientific applications-related issues. The course covers a wide range of topics, including the approximation of functions and integrals and the approximate solution of algebraic, transcendental, differential, and integral equations, with a focus on the stability, accuracy, efficiency, and dependability of numerical algorithms.

Numerical methods are the primary focus of studies due to the ease with which they may be implemented to solve different types of differential equations using a programmable approach. With the development of technology, there is an abundance of mathematical software that supports researchers in discovering numerical solutions to differential equations by programming the procedure under consideration. A great deal of progress has been made in the field of numerical analysis; yet, there is no method capable of solving all types of differential equations. As a result, academics are consistently putting up attempts to invent new numerical schemes. The objective of this book is to present the sophisticated numerical approaches that have been established in the field of computational techniques through original research in recent years. The book covers a broad spectrum of numerical issues that are currently implemented by the researchers to solve the associate real-life applications.

List of Figures

Figure 1.1	A numerical representation for eqn (1.6) in three different graphs' style.	8
Figure 1.2	A numerical representation for eqn (1.8) in three different graphs' style.	9
Figure 1.3	A numerical representation for eqn (1.10) in three different graphs' style.	10
Figure 1.4	A numerical representation for eqn (1.12) in three different graphs' style.	10
Figure 1.5	A numerical representation for eqn (1.13) in three different graphs' style.	11
Figure 1.6	A numerical representation for eqn (1.14) in three different graphs' style.	11
Figure 1.7	Pulse propagation in optical fibers with regard to dispersion.	12
Figure 1.8	The dispersion effect in optical fibers regarding pulse propagation.	13
Figure 2.1	At the top and bottom, absolute errors ($t = 5$) and physical behaviour ($t \leq 5$) plots of nHCB-DQM solutions, respectively, for Example 2.1.	28
Figure 2.2	At the top and bottom, absolute errors ($t = 5$) and physical behavior ($t \leq 5$) plots of nHCB-DQM solutions, respectively, for Example 2.2.	29
Figure 2.3	At the top and bottom, absolute errors ($t = 5$) and physical behaviour ($t \leq 5$) plots of nHCB-DQM solutions, respectively, for Example 2.3.	31
Figure 4.1	Graphical representation to the Example 1.	57
Figure 4.2	Graphical representation to the Example 5.	58
Figure 4.3	Graphical representation to the Example 6.	59
Figure 6.1	Absolute errors for Example 1 at $t \leq 0.5$ obtained by MCB-DQM and CFDS6.	93

Figure 6.2	Physical behavior of the solution of Example 1 at $t \leq 0.5$.	94
Figure 6.3	Absolute errors for Example 2 at $t \leq 5$ obtained by MCB-DQM and CFDS6.	95
Figure 6.4	Physical behavior of the solution of Example 2 at $t \leq 5$.	95
Figure 6.5	Physical behavior of the solution of Example 3 at $t \leq 10$.	96
Figure 7.1	Compatibility of u and solution profiles at $t = 0.1$ for $N = 11$ for Example 1.	113
Figure 7.2	Compatibility of u and v solution profiles at $t = 0.5$ for $N = 51$ for Example 1.	113
Figure 8.1	Numerical simulations of eqn (8.4) in distinct graphs' type.	126
Figure 8.2	Numerical simulations of eqn (8.5) in distinct graphs' type.	126
Figure 8.3	Numerical simulations of eqn (8.7) in distinct graphs' type.	127
Figure 8.4	Numerical simulations of eqn (8.8) in distinct graphs' type.	127
Figure 8.5	Matching between computational and numerical solutions based on Table 8.1.	128
Figure 8.6	In a crystal, vibrations are caused by waves passing through its atoms.	128
Figure 8.7	An atom in a crystal vibrates as waves pass through it.	129
Figure 9.1	Comparison of solution of Example 1 for different values of α.	144
Figure 9.2	Comparison of solution of Example 1 for different time levels at $\alpha = 0.4$.	144

List of Tables

Table 2.1	Values of $nH_i^4(x)$, $\frac{dnH_i^4}{dx}$, and $\frac{d^2nH_i^4}{dx^2}$ at x_j	24
Table 2.2	The absolute L_2 and L_∞ errors at $h = 0.01$ for example 2.1.	28
Table 2.3	The absolute L_2 and L_∞ errors at $h = 0.01$ for Example 2.2.	30
Table 2.4	Absolute error for Example 2.3, with $\mu = 3.5$ and $h = 0.016667$ at $t = 1, 3, 5$.	30
Table 4.1	Comparison of numerical values from the Bessel collocation method and B-spline method in terms of absolute error.	57
Table 4.2	Comparison of exact and numerical values at node points.	58
Table 4.3	Comparison of exact and numerical values at node points.	59
Table 5.1	At the nodal points, values of $B_i(w)$ for cubic B-spline and its derivatives.	66
Table 5.2	At the nodal points, value of $\boldsymbol{B_i}(\boldsymbol{w})$ for quartic B-spline and its derivatives.	67
Table 5.3	At the nodal points, value of $B_i(w)$ for quintic B-spline and its derivatives.	67
Table 5.4	Exponential B-spline values.	73
Table 5.5	Values of $B_i(x)$ and its first and second derivatives at the knot points.	74
Table 5.6	$T_m(w)$ and its derivatives at connecting points.	74
Table 5.7	Table for the values of UAH tension B-spline of order 4, i.e., UAH$B_{(i,4)}$ (w) and UAH$B'_{(i,4)}$ (w) at different node points.	76
Table 5.8	Value of UAT tension B-spline at different node points.	77
Table 5.9	$T_{m,5}(w)$ and its derivatives at connecting points.	79

List of Tables

Table 6.1	Absolute errors via MCB-DQM and CFDS6 for Example 1 at time t, $t \leq 0.4$.	92
Table 6.2	Thousand times of the absolute errors via CFDS6, MCB-DQM, and TMM for Example 1 at $t = 0.5$.	92
Table 6.3	L_∞ and L_2 errors for Example 1 via MCB-DQM and CFDS6 at time t, $t \leq 0.5$.	93
Table 6.4	Comparison for Example 2 taking $h = 1/60$	94
Table 6.5	Comparison in Example 2 for MCB-DQM, CFDS6, and RBF for $h = 0.02$.	95
Table 6.6	Comparison of errors and computational time in Example 3 for MCB-DQM, CFDS6, and RBF.	96
Table 6.7	Comparison in Example 3 for MCB-DQM, CFDS6, and TMM.	97
Table 8.1	Analytical and approximate values for $x \in \{-30, -29, -28, ..., -2, -1, 0\}$.	125

List of Contributors

Arora, Geeta, *Department of Mathematics, School of Chemical Engineering and Physical Sciences, Lovely Professional University, India*

Arora, Shelly, *Department of Mathematics, Panjabi University, India*

Attia, Raghda A. M., *Department of Basic Science, Higher Technological Institute 10th of Ramadan City, Egypt*

Bala, Indu, *Department of Mathematics, Panjabi University, India*

Cattani, C., *Engineering School, DEIM, University of Tuscia, P.le dell'Universit'a, Italy*

Emadifar, Homan, *Department of Mathematics, Hamedan Branch, Islamic Azad University, Iran*

Günerhan, Hatıra, *Department of Mathematics, Faculty of Education, Kafkas University, Turkey*

Gupta, M., *Department of Mathematics, School of Physical and Decision Sciences, Babasaheb Bhimrao Ambedkar University, India*

Kapoor, Mamta, *Department of Mathematics, Lovely Professional University, India*

Kaur, Ravinder, *Lovely Professional University, India*

Khademi, Masoumeh, *Department of Mathematics, Hamedan Branch, Islamic Azad University, Iran*

Khater, Mostafa M. A., *School of Medical Informatics and Engineering, Xuzhou Medical University, China; Department of Basic Science, Obour High Institute for Engineering and Technology, Egypt*

Kumar, Rakesh, *Lovely Professional University, India*

Mishra, Shubham, *Department of Mathematics, Lovely Professional University, India*

Pant, Rajendra, *Department of Mathematics, School of Chemical Engineering and Physical Sciences, Lovely Professional University, India*

Rakhra, Manik, *School of Computer Science and Engineering, Lovely Professional University, India*

Shukla, J. P., *School of Physical and Decision Sciences, Department of Mathematics, Babasaheb Bhimrao Ambedkar University, India*

Singh, B. K., *Department of Mathematics, School of Physical and Decision Sciences, Babasaheb Bhimrao Ambedkar University, India*

Singh, Neetu, *Department of Applied Sciences and Humanities, KCNIT, India*

List of Abbreviations

ADM	Adomian decomposition method
BC	Boundary condition
BCM	Bessel collocation method
CBCM	Cubic B-spline collocation method
CCP	Capacitively coupled plasmas
C-D	Convection–Diffusion
CFDS	Sixth-order compact finite difference scheme
CIR	Committed information rate
CVD	Chemical vapor deposition
DQM	Differential quadrature method
EW	Equal width
FDS	Finite difference scheme
FEC	Forward error correction
FFCH	Fuchssteiner–Fokas–Camassa–Holm
FLFPP	Fuzzy linear fractional programming problem
FLPP	Fractional linear programming problem
FPP	Fractional programming problem
FW	Fornberg–Whitham
GP	Gilson–Pickering
HAM	Homotopy Analysis method
HTT	Hyperbolic–trigonometric tension
IC	Initial condition
ICP	Inductively Coupled Plasma
KdV	Korteweg–de Vries
KG	Klein–Gordon
LFP	Linear fractional programming
LFPP	Linear fractional programming problem
LPP	Linear programming problem
MHD	Magneto-Hydrodynamic
m-KdV	Modified Korteweg–de Vries
m-RLW	Modified regularized long wave

nHCB–DQM	New hybrid cubic B–spline differential quadrature method
ODE	Ordinary differential equation
PDE	Partial differential equation
PDQM	Polynomial DQM
QTT	Quartic trigonometric
RBF	Radial basis function
RCBCM	Redefined CBCM
RH	Rosenau–Hyman
RK	Runge Kutta
RPSM	Residual power series method
RR	Rangwala–Rao
SSP	Strong stability preserving
TIR	Total internal reflection
TMM	Taylor matrix method
TrpFN	Trapezoidal fuzzy number
UAH	Uniform algebraic hyperbolic
UAT	Uniform algebraic trigonometric
VIT	Variational iteration

1

A Slow Varying Envelope of the Electric Field is Influenced by Integrability Conditions

Mostafa M. A. Khater[1,2]

[1]School of Medical Informatics and Engineering,
Xuzhou Medical University, China
[2]Department of Basic Science,
Obour High Institute for Engineering and Technology, Egypt
Email: mostafa.khater2024@yahoo.com

Abstract

In this study, we explore how the integrability condition affects nonlinear Schrödinger equations with mixed derivatives. These effects were first formulated mathematically in 1990 by A. Rangwala, who gave his formulation the name Rangwala–Rao ($\mathcal{RR}$) equation. We are looking at new soliton wave solutions and how they interact in order to get a clear picture of the slowly changing envelope of the electric field and pulse propagation in optical fibers when it comes to the dispersion effect. Modern computational methods like the well-known Khater II method and Sardar Sub-equation method are used to find original solitary wave solutions for the model being studied. Numbers calculate these solutions to show how dynamic optical fiber pulse propagation is. We demonstrate the paper's novelty by contrasting our findings with those of other researchers.

Keywords: electric field, Rangwala–Rao equation, Khater II method, solitary wave.

AMS classification: 35Q60, 35E05, 35C08, 35Q51.

1.1 Introduction

To make an optical fiber, glass or plastic is stretched to the size of a human hair. Optical fibers can transport data over more considerable distances and faster than electrical cables. Fibers lose less signal and are not affected by electromagnetic fields. Fibers in a fiberscope transport light into or images out of a tiny space. Specially designed fibers are used in numerous ways, including sensors and lasers. Optical fibers are clear, low-index-of-refraction material wrapped around a core. CIR helps the fiber act like a waveguide, trapping light. Multi-mode fibers may have numerous transverse propagation modes. Multi-mode fibers' larger cores make them ideal for sending a lot of power or communicating over short distances. Most communication lines use multi-mode fibers, not single-mode [1, 2].

Reliable optical fiber transmission requires low signal deterioration while connecting fibers. This involves perfect fiber cleaving, alignment, and connection, unlike joining electrical wire or cable. Fusion splices are used to create permanent connections. An electric arc melts the fiber ends together. Mechanical splicing presses fiber ends together. Specialized connections join optical fibers temporarily or semi-permanently. Fiber optics involve designing and utilizing optical fibers. Physicist Narinder Singh Kapany initially used [3, 4].

Daniel Colladon and Jacques Babinet initially demonstrated fiber optics in Paris in the 1840s. John Tyndall showed it in London 12 years later. Tyndall mentioned internal reflection in his 1870 book on light. Nineteenth century Viennese doctors bent glass rods to illuminate patients' bodies. In the early 20th century, it was used to decorate dentists' offices. Clarence Hansell and John Logie Baird separately achieved picture transmission using tubes in the 1920s. Heinrich Lamm proved in the 1930s that unclad optical fibers might convey images for medical checks, but his work has been mostly forgotten. Harold Hopkins and Narinder Singh Kapany of Imperial College sent a photograph of over 10,000 fibers and 3000 fibers 75 cm long in the same year. Basil Hirschowitz, C. Wilbur Peters, and Lawrence E. Curtiss created the first fiber optic semi-flexible gastroscope in 1956. Prior optical fibers used air or unusable oils and waxes as cladding. Curtiss invented the gastroscope using glass-clad fibers [5, 6, 7].

Manfred Börner showed the first fiber-optic data transmission system in 1965 at Telefunken Research Labs in Ulm. The first patent was submitted in 1966. 1968 NASA moon cameras used fiber optics. Workers operating the cameras must be supervised by someone with the proper clearance. In 1965, Charles K. Kao and George A. Hockham of Standard Telephones and Cables suggested decreasing optical fiber attenuation to 20 dB per kilometer, making fibers a feasible communication medium. Instead of basic physical causes like scattering, they blamed contaminants that could be removed. They properly predicted optical fiber light loss and suggested using highly pure silica glass. Kao, winner of 2009 Nobel Prize in Physics, made this discovery. Researchers at Corning Glass Works were the first to attain 20 dB/km in 1970. Titanium added to silica glass might provide a 17-dB/km fiber. A few years later, they created a fiber with 4 dB/km attenuation using germanium dioxide. General Electric initially made large strands of fused quartz nuggets in 1981 [8, 9].

Initially, quality optical fibers were manufactured at 2 m/s. Thomas Mensah, a chemical engineer at Corning, increased the manufacturing speed to 50 m/s, decreasing the cost of optical fiber cables compared to copper. These advancements ushered in the era of optical fiber for phones and other communications. The first metropolitan fiber optic cable was erected in Turin, Italy, in 1977, courtesy of CSELT and Corning's work on practical optical fiber cables. CSELT also invented Springroof optical fiber splicing. The low attenuation of contemporary optical cables allows for long-distance fiber connections. David N. Payne of Southampton and Emmanuel Desurvire of Bell Labs created the erbium-doped fiber amplifier in 1986 and 1987. This invention reduced the cost of long-distance fiber networks by eliminating the requirement for optical-electrical-optical repeaters [10, 11, 12].

Photonic-crystal fiber, which guides light by diffraction rather than TIR, was created in 1991. Businesses got photonic crystal fibers in 2000. Photonic crystal fibers may transfer more power than standard fibers and have wavelength-dependent properties. Optical fiber is used in telecommunications and computer networking because it is flexible and may be bundled into cables. Because infrared light attenuates less than electricity through wires, it is useful for long-distance communications. This allows long-distance communication with a few repeaters. Thanks to wavelength-division multiplexing, a fiber may carry many channels at once. Net data rate per cable equals channel data rate less FEC overhead times channels. Short-distance applications like an office network may benefit from fiber-optic wiring. A single fiber can transfer more data than category 5 cable, which works at 100 Mbit/s or 1 Gbit/s [13, 14, 15].

Short-distance device connections employ fiber optics. Most current HDTVs include digital audio optical links. S/PDIF over TOSLINK is possible. Remote sensing uses fibers to make antennas. The optical fiber sensor has several uses. Fiber sends and measures radiation. Linking a sensor to an analytical tool often requires a fiber optic connection [16]. By modifying a fiber's intensity, phase, polarization, wavelength, or transit time, they may be used as strain, temperature, and pressure sensors. Light intensity sensors need a light source and a detector. Fiber optic sensors may offer scattered sensing up to 1 m away. Distributed acoustic sensors are one technique [17, 18].

Smaller sensor devices on the fiber tip may provide accurate measurements. These may be manufactured utilizing microfabrication and nanofabrication processes, remaining within the fiber tip's minuscule limit for usage in a blood artery. In extrinsic fiber optic sensors, modulated light is delivered from a non-fiber optical sensor or electronic sensor to an optical transmitter using a multi-mode optical fiber cable. Extrinsic sensors can be utilized when conventional ones cannot. A pyrometer installed outside a jet engine can read its inside temperature owing to a radiation-carrying fiber. When electromagnetic fields in electrical transformers make other measures unusable, extrinsic sensors may acquire reliable interior temperature readings. Outside sensors measure vibration, displacement, speed, acceleration, torque, and torsion. The solid-state gyroscope uses light interference. The fiber optic gyroscope can detect mechanical rotation via the Sagnac effect [19, 20, 21].

Fiber optic sensors are used in high-tech security systems to detect intruders. The signal from a fiber optic sensor wire mounted on a fence, pipeline, or communication cable is monitored for irregularities. The returning signal is analyzed digitally to detect an intruder. Optical fibers are important in optical, chemical, and biosensors. Solar cells and optical fibers can transmit energy across large distances. This power transfer method is not as successful as others, but it is useful near MRI machines, where a metallic conductor would be detrimental. Two further uses are high-powered antenna elements and high-voltage transmission measurement tools [22, 23, 24, 25].

Recently, many optical phenomena have been represented through nonlinear partial differential equation. Additionally, some accurate and novel computational schemes have been formulated to investigate the analytical and numerical solutions of the formulated model [26, 27]. In this context, we are studying the $\mathcal{RR}$ equation, which is given by [28, 29, 30]

$$\mathcal{U}_{xt} - \rho_1 \mathcal{U}_{xx} + \mathcal{U} + i\rho_2 |\mathcal{U}|^2 \mathcal{U}_x = 0, \qquad (1.1)$$

where $\mathcal{U} = \mathcal{U}(x,t)$ represents a complex smooth envelope function, while ρ_1 and ρ_2 are real constants. Using the next transformation $\mathcal{U}(x,t) = e^{i\mathfrak{Q}(\mathfrak{Z})}, e^{-i,t,\vartheta}, \mathfrak{T}(\mathfrak{Z}), \mathfrak{Z} = x - \varepsilon t$ to eqn (1.1) gets

$$\begin{cases} \text{Re: } (-\varepsilon - \rho_1)\mathfrak{T}'' + \mathfrak{T}(\varepsilon + \rho_1)(\mathfrak{Q}')^2 - \rho_2 \mathfrak{T}^3 \mathfrak{Q}' + \mathfrak{T}\vartheta \mathfrak{Q}' + \mathfrak{T} = 0, \\ \text{Im: } -\mathfrak{T}'(2(\varepsilon + \rho_1)\mathfrak{Q}' - \rho_2 \mathfrak{T}^2 + \vartheta) - \mathfrak{T}(\varepsilon + \rho_1)\mathfrak{Q}'' = 0. \end{cases} \quad (1.2)$$

Substituting $\mathfrak{Q} = \int \frac{(\rho_2 \mathfrak{T}(x)^2 - \vartheta)^2}{4\rho_2(\varepsilon + \rho_1)\mathfrak{T}(x)^2} \, dx$ into the first equation of (1.2) leads to

$$(-16\varepsilon^2 \rho_2^2 - 32\varepsilon \rho_1 \rho_2^2 - 16\rho_1^2 \rho_2^2)\mathfrak{T}^3 \mathfrak{T}'' + \mathfrak{T}^4 (16\varepsilon \rho_2^2 - 6\rho_2^2 \vartheta^2 + 16\rho_1 \rho_2^2)$$
$$+ 8\rho_2^3 \vartheta \mathfrak{T}^6 - 3\rho_2^4 \mathfrak{T}^8 + \vartheta^4 = 0. \quad (1.3)$$

Balancing the terms of eqn (1.3) by using the homogeneous balance rule and the suggested computational schemes' auxiliary equations $\Big[f'(\mathfrak{Z}) \to -\delta - f(\mathfrak{Z})^2, \phi'(\mathfrak{Z}) \to -f(\mathfrak{Z})\phi(\mathfrak{Z})\, \&\, \phi'(\mathfrak{Z})^2 \to \gamma_3\, \phi(\mathfrak{Z})^4 + \gamma_2\, \phi(\mathfrak{Z})^3 + \gamma_1\, \phi(\mathfrak{Z})^2\Big]$, where $\delta, \gamma_1, \gamma_2,$ and γ_3 are arbitrary constants, leads to $N = \frac{1}{2}$. Consequently, we have to use another transformation that is given by $\mathfrak{T}(\mathfrak{Z}) = \sqrt{\mathcal{V}(\mathfrak{Z})}$. Thus, eqn (1.3) transforms to be in the following formula:

$$\mathcal{V}\mathcal{V}'' + \mathcal{V}^4 \varsigma_2 - \mathcal{V}^3 \varsigma_3 - \mathcal{V}^2 \varsigma_4 - \varsigma_1 (\mathcal{V}')^2 + \varsigma_5 = 0, \quad (1.4)$$

where $\Big[\varsigma_1 = \frac{1}{2}, \varsigma_2 = \frac{3\rho_2^2}{8(\varepsilon+\rho_1)^2}, \varsigma_3 = \frac{\rho_2 \vartheta}{(\varepsilon+\rho_1)^2}, \varsigma_4 = \frac{8\varepsilon + 8\rho_1 - 3\vartheta^2}{4(\varepsilon+\rho_1)^2}, \varsigma_5 = -\frac{\vartheta^4}{8\rho_2^2(\varepsilon+\rho_1)^2}\Big]$. Using the homogeneous balance rule on eqn (1.4) leads to $N = 1$. Thus, the general solutions of the suggested model through the above-mentioned computational schemes are formulated by

$$\mathcal{V}(\mathfrak{Z}) = \begin{cases} \sum_{i=1}^{n} \left(a_i f(\mathfrak{Z})^i + b_i \phi(\mathfrak{Z}) f(\mathfrak{Z})^{i-1}\right) + a_0 = a_1 f(\mathfrak{Z}) + a_0 + b_1 \phi(\mathfrak{Z}), \\ \\ \sum_{i=1}^{n} \left(a_i \phi(\mathfrak{Z})^i + \frac{b_i}{\phi(\mathfrak{Z})^i}\right) + a_0 = a_1 \phi(\mathfrak{Z}) + a_0 + \frac{b_1}{\phi(\mathfrak{Z})}, \end{cases} \quad (1.5)$$

where $a_0, a_1,$ and b_1 are arbitrary constants to be evaluated through the methods' frameworks.

6 A Slow Varying Envelope of the Electric Field is Influenced by Integrability

The remaining sections of the chapter are organized as follows. Section 1.2 explains some novel solitary wave solutions of the $\mathcal{RR}$ equation. Section 1.3 explains the study's scientific contributions and results' novelty. Section 1.4 gives the conclusion of the whole study.

1.2 Solitary Wave Solutions

Here, we apply two accurate and recent computational schemes to the $\mathcal{RR}$ model for constructing some novel solitary wave solutions. These solutions are illustrated by using some distinct graphs' style.

1.2.1 The Khater II method's results

Investigating the above-parameters' values through the Khater II method's framework obtains

Set I

$$a_0 \to \frac{2\delta - \varsigma_4}{\varsigma_3}, a_1 \to \frac{i(\varsigma_4 - 2\delta)}{\sqrt{\delta\varsigma_3}}, b_1 \to 0, \varsigma_1 \to \frac{\varsigma_4}{4\delta} + 1, \varsigma_2 \to \frac{\varsigma_3^2(4\delta - \varsigma_4)}{4(\varsigma_4 - 2\delta)^2}.$$

Set II

$$a_0 \to \frac{\varsigma_3^2 - \sqrt{\varsigma_3^4 + 4\varsigma_2\varsigma_4\varsigma_3^2}}{4\varsigma_2\varsigma_3}, a_1 \to \sqrt{2}\sqrt{-\frac{\varsigma_4}{\varsigma_3^2 + \sqrt{\varsigma_3^4 + 4\varsigma_2\varsigma_4\varsigma_3^2} + 4\varsigma_2\varsigma_4}},$$

$$b_1 \to 0, \delta \to \frac{\varsigma_3^2 - \sqrt{\varsigma_3^4 + 4\varsigma_2\varsigma_4\varsigma_3^2} + 4\varsigma_2\varsigma_4}{8\varsigma_2},$$

$$\varsigma_1 \to \frac{\varsigma_3^2}{2\sqrt{\varsigma_3^4 + 4\varsigma_2\varsigma_4\varsigma_3^2}} + \frac{3}{2}.$$

Set III

$$a_0 \to \frac{\varsigma_3^2 - \sqrt{\varsigma_3^4 + 4\varsigma_2\varsigma_4\varsigma_3^2}}{4\varsigma_2\varsigma_3}, a_1 \to \frac{\sqrt{\frac{\varsigma_4\left(-\varsigma_3^2 + \sqrt{\varsigma_3^4 + 4\varsigma_2\varsigma_4\varsigma_3^2} - 4\varsigma_2\varsigma_4\right)}{\varsigma_2}}}{2\sqrt{2}\sqrt{\varsigma_4\left(\varsigma_3^2 + 4\varsigma_2\varsigma_4\right)}},$$

$$b_1 \to \frac{\sqrt{\frac{-\varsigma_3^2+\sqrt{\varsigma_3^4+4\varsigma_2\varsigma_4\varsigma_3^2}-2\varsigma_2\varsigma_4}{\varsigma_2^2}}}{2\sqrt{2}},$$

$$\delta \to \frac{\varsigma_3^2-\sqrt{\varsigma_3^4+4\varsigma_2\varsigma_4\varsigma_3^2}+4\varsigma_2\varsigma_4}{2\varsigma_2}, \varsigma_1 \to \frac{\varsigma_3^2}{2\sqrt{\varsigma_3^4+4\varsigma_2\varsigma_4\varsigma_3^2}}+\frac{3}{2}.$$

Thus, the solitary wave solutions of the investigated model are given by For $\delta \neq 0$, we get

$$\mathscr{U}_{\mathrm{I},1}(x,t) = e^{i\int \frac{(\rho_2 \mathfrak{T}(x)^2-\vartheta)^2}{4\rho_2(\varepsilon+\rho_1)\mathfrak{T}(x)^2}\,dx}$$
$$e^{-it\vartheta}\sqrt{\frac{(2\delta-\varsigma_4)}{\varsigma_3}\left(1+i\tan\left(\sqrt{\delta}(x-\varepsilon t)\right)\right)}, \quad (1.6)$$

$$\mathscr{U}_{\mathrm{I},2}(x,t) = e^{i\int \frac{(\rho_2 \mathfrak{T}(x)^2-\vartheta)^2}{4\rho_2(\varepsilon+\rho_1)\mathfrak{T}(x)^2}\,dx}$$
$$e^{-it\vartheta}\sqrt{\frac{i(\varsigma_4-2\delta)}{\varsigma_3}\left(\cot\left(\sqrt{\delta}(x-\varepsilon t)\right)+i\right)}, \quad (1.7)$$

$$\mathscr{U}_{\mathrm{II},1}(x,t) = \frac{1}{2}e^{i\int \frac{(\rho_2\mathfrak{T}(x)^2-\vartheta)^2}{4\rho_2(\varepsilon+\rho_1)\mathfrak{T}(x)^2}\,dx}\,e^{-it\vartheta}$$
$$\times\sqrt{\frac{\varsigma_3-\frac{\sqrt{\varsigma_3^4+4\varsigma_2\varsigma_4\varsigma_3^2}}{\varsigma_3}}{\varsigma_2}-4\sqrt{2}\sqrt{\delta}\sqrt{-\frac{\varsigma_4}{\varsigma_3^2+\sqrt{\varsigma_3^4+4\varsigma_2\varsigma_4\varsigma_3^2}+4\varsigma_2\varsigma_4}}\tan\left(\sqrt{\delta}(x-\varepsilon t)\right)}, \quad (1.8)$$

$$\mathscr{U}_{\mathrm{II},2}(x,t) = e^{i\int \frac{(\rho_2\mathfrak{T}(x)^2-\vartheta)^2}{4\rho_2(\varepsilon+\rho_1)\mathfrak{T}(x)^2}\,dx}\,e^{-it\vartheta}$$
$$\times\sqrt{\sqrt{2}\sqrt{\delta}\sqrt{-\frac{\varsigma_4}{\varsigma_3^2+\sqrt{\varsigma_3^4+4\varsigma_2\varsigma_4\varsigma_3^2}+4\varsigma_2\varsigma_4}}\cot\left(\sqrt{\delta}(x-\varepsilon t)\right)+\frac{\varsigma_3^2-\sqrt{\varsigma_3^4+4\varsigma_2\varsigma_4\varsigma_3^2}}{4\varsigma_2\varsigma_3}}, \quad (1.9)$$

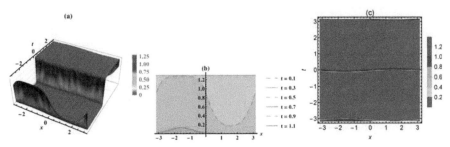

Figure 1.1 A numerical representation for eqn (1.6) in three different graphs' style.

$$\mathcal{U}_{\mathrm{III},1}(x,t) = \frac{1}{2} e^{i \int \frac{\left(\rho_2 \mathfrak{T}(x)^2 - \vartheta\right)^2}{4\rho_2 (\varepsilon + \rho_1) \mathfrak{T}(x)^2} dx} e^{-it\vartheta}$$

$$\times \left(\sqrt{2} \left(\sqrt{\frac{-\varsigma_3^2 + \sqrt{\varsigma_3^4 + 4\varsigma_2 \varsigma_4 \varsigma_3^2} - 2\varsigma_2 \varsigma_4}{\varsigma_2^2}} \sec\left(\sqrt{\delta}(x - \varepsilon t)\right) \right. \right.$$

$$\left. - \frac{\sqrt{\delta} \sqrt{\frac{\varsigma_4 \left(-\varsigma_3^2 + \sqrt{\varsigma_3^4 + 4\varsigma_2 \varsigma_4 \varsigma_3^2} - 4\varsigma_2 \varsigma_4\right)}{\varsigma_2}}}{\sqrt{\varsigma_4 \left(\varsigma_3^2 + 4\varsigma_2 \varsigma_4\right)}} \tan\left(\sqrt{\delta}(x - \varepsilon t)\right) \right) \tag{1.10}$$

$$+ \left. \frac{\varsigma_3 - \frac{\sqrt{\varsigma_3^4 + 4\varsigma_2 \varsigma_4 \varsigma_3^2}}{\varsigma_3}}{\varsigma_2} \right)^{\frac{1}{2}},$$

$$\mathcal{U}_{\mathrm{III},2}(x,t) = \frac{1}{2} e^{i \int \frac{\left(\rho_2 \mathfrak{T}(x)^2 - \vartheta\right)^2}{4\rho_2 (\varepsilon + \rho_1) \mathfrak{T}(x)^2} dx} e^{-it\vartheta}$$

$$\times \left(\sqrt{2} \left(\frac{\sqrt{\delta} \sqrt{\frac{\varsigma_4 \left(-\varsigma_3^2 + \sqrt{\varsigma_3^4 + 4\varsigma_2 \varsigma_4 \varsigma_3^2} - 4\varsigma_2 \varsigma_4\right)}{\varsigma_2}}}{\sqrt{\varsigma_4 \left(\varsigma_3^2 + 4\varsigma_2 \varsigma_4\right)}} \cot\left(\sqrt{\delta}(x - \varepsilon t)\right) \right. \right.$$

$$\left. + \sqrt{\frac{-\varsigma_3^2 + \sqrt{\varsigma_3^4 + 4\varsigma_2 \varsigma_4 \varsigma_3^2} - 2\varsigma_2 \varsigma_4}{\varsigma_2^2}} \csc\left(\sqrt{\delta}(x - \varepsilon t)\right) \right) \tag{1.11}$$

$$+ \left. \frac{\varsigma_3 - \frac{\sqrt{\varsigma_3^4 + 4\varsigma_2 \varsigma_4 \varsigma_3^2}}{\varsigma_3}}{\varsigma_2} \right)^{\frac{1}{2}}.$$

1.2.2 The Sardar sub-equation method's results

Investigating the above-parameters' values through the Sardar sub-equation method's framework leads to

Set I

$$a_1 \to \frac{\sqrt{a_0^2\left(9\gamma_2^2 - 32\gamma_1\gamma_3\right)} + 3a_0\gamma_2}{4\gamma_1}, b_1 \to 0, \varsigma_1 \to \frac{a_0\varsigma_2\left(3\gamma_2\sqrt{a_0^2\left(9\gamma_2^2 - 32\gamma_1\gamma_3\right)} + a_0\left(9\gamma_2^2 - 16\gamma_1\gamma_3\right)\right)}{8\gamma_1^2\gamma_3} + 2,$$

$$\varsigma_3 \to \frac{2a_0^3\left(16\gamma_1\gamma_3 - 3\gamma_2^2\right)\varsigma_2 - 2a_0^2\gamma_2\varsigma_2\sqrt{a_0^2\left(9\gamma_2^2 - 32\gamma_1\gamma_3\right)} + a_0\gamma_1\left(3\gamma_2^2 - 8\gamma_1\gamma_3\right) + (-\gamma_1)\gamma_2\sqrt{a_0^2\left(9\gamma_2^2 - 32\gamma_1\gamma_3\right)}}{8a_0^2\gamma_1\gamma_3},$$

$$\varsigma_4 \to \frac{2a_0^3\left(9\gamma_2^2 - 32\gamma_1\gamma_3\right)\varsigma_2 + 6a_0^2\gamma_2\varsigma_2\sqrt{a_0^2\left(9\gamma_2^2 - 32\gamma_1\gamma_3\right)} + a_0\gamma_1\left(32\gamma_1\gamma_3 - 9\gamma_2^2\right) + 3\gamma_1\gamma_2\sqrt{a_0^2\left(9\gamma_2^2 - 32\gamma_1\gamma_3\right)}}{16a_0\gamma_1\gamma_3},$$

$$\varsigma_5 \to \frac{a_0\left(2a_0^3\left(3\gamma_2^2 - 8\gamma_1\gamma_3\right)\varsigma_2 + 2a_0^2\gamma_2\varsigma_2\sqrt{a_0^2\left(9\gamma_2^2 - 32\gamma_1\gamma_3\right)} + a_0\gamma_1\left(16\gamma_1\gamma_3 - 3\gamma_2^2\right) + \gamma_1\gamma_2\sqrt{a_0^2\left(9\gamma_2^2 - 32\gamma_1\gamma_3\right)}\right)}{16\gamma_1\gamma_3}.$$

Set II

$$a_1 \to \frac{i\sqrt{\frac{3}{2}}\sqrt{\gamma_3}}{\sqrt{\varsigma_2}}, b_1 \to 0, \varsigma_1 \to \frac{1}{2}, \varsigma_3 \to \frac{8a_0\varsigma_2}{3} - \frac{i\sqrt{\frac{2}{3}}\gamma_2\sqrt{\varsigma_2}}{\sqrt{\gamma_3}},$$

$$\varsigma_4 \to \frac{1}{2}\left(\frac{i\sqrt{6}a_0\gamma_2\sqrt{\varsigma_2}}{\sqrt{\gamma_3}} - 4a_0^2\varsigma_2 + \gamma_1\right),$$

$$\varsigma_5 \to \frac{1}{6}a_0^2\left(\frac{i\sqrt{6}a_0\gamma_2\sqrt{\varsigma_2}}{\sqrt{\gamma_3}} - 2a_0^2\varsigma_2 + 3\gamma_1\right).$$

Set III

$$a_0 \to \frac{i\sqrt{\frac{3}{2}}\gamma_2}{2\sqrt{\gamma_3}\sqrt{\varsigma_2}}, a_1 \to \frac{i\sqrt{\frac{3}{2}}\sqrt{\gamma_3}}{\sqrt{\varsigma_2}}, b_1 \to 0, \varsigma_1 \to \frac{1}{2}, \varsigma_3 \to \frac{i\sqrt{\frac{2}{3}}\gamma_2\sqrt{\varsigma_2}}{\sqrt{\gamma_3}},$$

$$\varsigma_4 \to \frac{\gamma_1}{2}, \varsigma_5 \to \frac{3\gamma_2^2\left(\gamma_2^2 - 4\gamma_1\gamma_3\right)}{64\gamma_3^2\varsigma_2}.$$

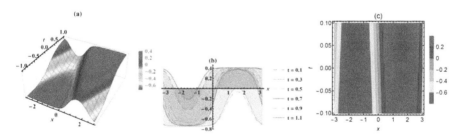

Figure 1.2 A numerical representation for eqn (1.8) in three different graphs' style.

10 *A Slow Varying Envelope of the Electric Field is Influenced by Integrability*

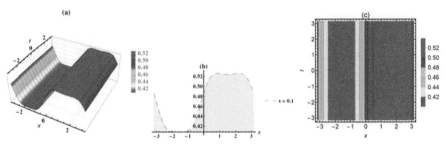

Figure 1.3 A numerical representation for eqn (1.10) in three different graphs' style.

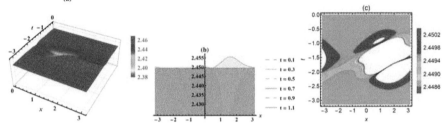

Figure 1.4 A numerical representation for eqn (1.12) in three different graphs' style.

Thus, the solitary wave solutions of the investigated model are given by the following. For $\gamma_1 \neq 0$, $\gamma_2 \neq 0$, and $\gamma_3 \neq 0$, we get

$$\mathscr{U}_{\mathrm{I},1}(x,t) = e^{i\int \frac{(\rho_2 \mathfrak{T}(x)^2 - \vartheta)^2}{4\rho_2(\varepsilon+\rho_1)\mathfrak{T}(x)^2} dx} e^{-it\vartheta}$$

$$\times \sqrt{\frac{\left(\sqrt{a_0^2\left(9\gamma_2^2 - 32\gamma_1\gamma_3\right)} + 3a_0\gamma_2\right)\left(\sqrt{(\gamma_2^2 - 4\gamma_1\gamma_3)}\tanh^2(\chi_1)(-\mathrm{sech}^2(\chi_1)) - \gamma_2\mathrm{sech}^2(\chi_1)\right)}{2\left(\gamma_2^2 - 4\gamma_1\gamma_3\tanh^2(\chi_1)\right)} + a_0},$$

(1.12)

$$\mathscr{U}_{\mathrm{II},1}(x,t) = e^{i\int \frac{(\rho_2 \mathfrak{T}(x)^2 - \vartheta)^2}{4\rho_2(\varepsilon+\rho_1)\mathfrak{T}(x)^2} dx} e^{-it\vartheta}$$

$$\times \sqrt{a_0 + \frac{i\sqrt{6}\sqrt{\gamma_3}\gamma_1\left(\sqrt{(\gamma_2^2 - 4\gamma_1\gamma_3)}\tanh^2(\chi_1)(-\mathrm{sech}^2(\chi_1)) - \gamma_2\mathrm{sech}^2(\chi_1)\right)}{\sqrt{\varsigma_2}\left(\gamma_2^2 - 4\gamma_1\gamma_3\tanh^2(\chi_1)\right)}},$$

(1.13)

$$\mathscr{U}_{\mathrm{III},1}(x,t) = e^{i\int \frac{(\rho_2 \mathfrak{T}(x)^2 - \vartheta)^2}{4\rho_2(\varepsilon+\rho_1)\mathfrak{T}(x)^2} dx} e^{-it\vartheta}$$

$$\times \frac{\sqrt[4]{3}\sqrt{-\frac{i\left(\gamma_2^3 - 4\gamma_1\gamma_3\gamma_2 + 4\gamma_1\gamma_3\sqrt{(\gamma_2^2 - 4\gamma_1\gamma_3)}\tanh^2(\chi_1)(-\mathrm{sech}^2(\chi_1))\right)}{\sqrt{\gamma_3}\sqrt{\varsigma_2}\left(4\gamma_1\gamma_3\tanh^2(\chi_1) - \gamma_2^2\right)}}}{2^{3/4}}$$

(1.14)

For $\gamma_1 = 0$, $\gamma_2 \neq 0$, and $\gamma_3 \neq 0$, we get

$$\mathscr{U}_{\mathrm{II},2}(x,t) = e^{i\int \frac{(\rho_2 \mathfrak{T}(x)^2 - \vartheta)^2}{4\rho_2(\varepsilon+\rho_1)\mathfrak{T}(x)^2} dx} e^{-it\vartheta} \times \sqrt{a_0 + \frac{2i\sqrt{6}\sqrt{\gamma_3}\gamma_2}{\sqrt{\varsigma_2}\left(\gamma_2^2(-\varepsilon t + x + \varsigma)^2 - 4\gamma_3\right)}}, \quad (1.15)$$

1.2 Solitary Wave Solutions 11

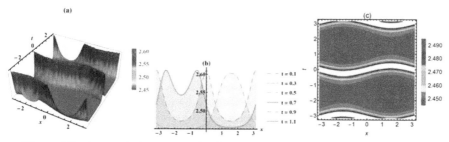

Figure 1.5 A numerical representation for eqn (1.13) in three different graphs' style.

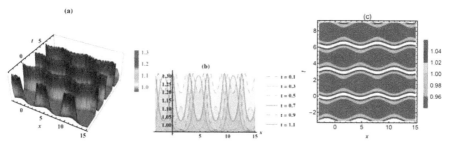

Figure 1.6 A numerical representation for eqn (1.14) in three different graphs' style.

$$\mathscr{U}_{\mathrm{III},2}(x,t) = e^{i\int \frac{\left(\rho_2 \mathfrak{T}(x)^2 - \vartheta\right)^2}{4\rho_2(\varepsilon+\rho_1)\mathfrak{T}(x)^2}\, dx}\, e^{-it\vartheta} \times \frac{\sqrt[4]{3}\sqrt{\frac{i\gamma_2\left(\frac{8\gamma_3}{\gamma_2^2(-\varepsilon t+x+\varsigma)^2-4\gamma_3}+1\right)}{\sqrt{\gamma_3}\sqrt{\varsigma_2}}}}{2^{3/4}}. \quad (1.16)$$

For $\gamma_1 \neq 0$, $\gamma_2 = 0$, and $\gamma_3 \neq 0$, we get

$$\mathscr{U}_{1,2}(x,t) = e^{i\int \frac{\left(\rho_2 \mathfrak{T}(x)^2 - \vartheta\right)^2}{4\rho_2(\varepsilon+\rho_1)\mathfrak{T}(x)^2}\, dx}\, e^{-it\vartheta} \times \sqrt{\frac{\sqrt{2}\sqrt{-a_0^2\gamma_1\gamma_3}\sqrt{-\operatorname{sech}^2(\chi_3)}}{\sqrt{\gamma_1}\sqrt{\gamma_3}} + a_0}, \quad (1.17)$$

$$\mathscr{U}_{\mathrm{II},3}(x,t) = e^{i\int \frac{\left(\rho_2 \mathfrak{T}(x)^2 - \vartheta\right)^2}{4\rho_2(\varepsilon+\rho_1)\mathfrak{T}(x)^2}\, dx}\, e^{-it\vartheta} \times \sqrt{a_0 + \frac{i\sqrt{\frac{3}{2}}\sqrt{\gamma_1}\sqrt{-\operatorname{sech}^2(\chi_3)}}{\sqrt{\varsigma_2}}}, \quad (1.18)$$

$$\mathscr{U}_{\mathrm{III},3}(x,t) = e^{i\int \frac{\left(\rho_2 \mathfrak{T}(x)^2 - \vartheta\right)^2}{4\rho_2(\varepsilon+\rho_1)\mathfrak{T}(x)^2}\, dx}\, e^{-it\vartheta} \times \sqrt[4]{\frac{3}{2}}\sqrt{\frac{i\sqrt{\gamma_1}\sqrt{-\operatorname{sech}^2(\chi_3)}}{\sqrt{\varsigma_2}}}. \quad (1.19)$$

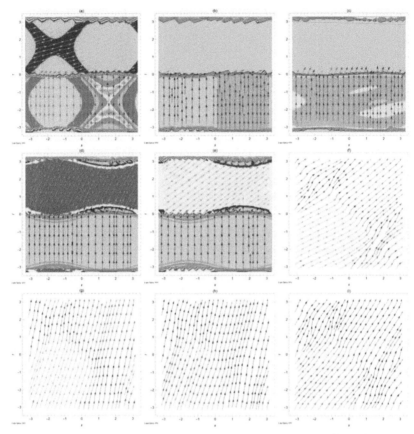

Figure 1.7 Pulse propagation in optical fibers with regard to dispersion.

For $\gamma_1 = \frac{\gamma_2^2}{4}$, and $\gamma_2 \neq 0$, $\gamma_3 \neq 0$, we get

$$\mathscr{U}_{1,3}(x,t) = e^{i\int \frac{(\rho_2 \mathfrak{T}(x)^2 - \vartheta)^2}{4\rho_2(\varepsilon+\rho_1)\mathfrak{T}(x)^2}dx} e^{-it\vartheta}$$

$$\times \sqrt{\frac{\left(\sqrt{a_0^2\left(9\gamma_2^2 - 8\gamma_2^2\gamma_3\right)} + 3a_0\gamma_2\right)\left(\gamma_2\operatorname{sech}^2\left(\frac{1}{2}\chi_2\right) + 4\sqrt{\gamma_2^2(\gamma_3-1)}\sinh^6\left(\frac{1}{2}\chi_2\right)\operatorname{csch}^4(\chi_2)\right)}{2\gamma_2^2\left(\gamma_3\tanh^2\left(\frac{1}{2}\chi_2\right) - 1\right)}} + a_0,$$

(1.20)

$$\mathscr{U}_{\mathrm{II},4}(x,t) = e^{i\int \frac{(\rho_2 \mathfrak{T}(x)^2 - \vartheta)^2}{4\rho_2(\varepsilon+\rho_1)\mathfrak{T}(x)^2}dx} e^{-it\vartheta}$$

$$\times \sqrt{a_0 + \frac{i\sqrt{\frac{3}{2}}\sqrt{\gamma_3}\left(\gamma_2\operatorname{sech}^2\left(\frac{1}{2}\chi_2\right) + 4\sqrt{\gamma_2^2(\gamma_3-1)}\sinh^6\left(\frac{1}{2}\chi_2\right)\operatorname{csch}^4(\chi_2)\right)}{\sqrt{\varsigma_2}\left(2\gamma_3\tanh^2\left(\frac{1}{2}\chi_2\right) - 2\right)}},$$

(1.21)

1.2 Solitary Wave Solutions 13

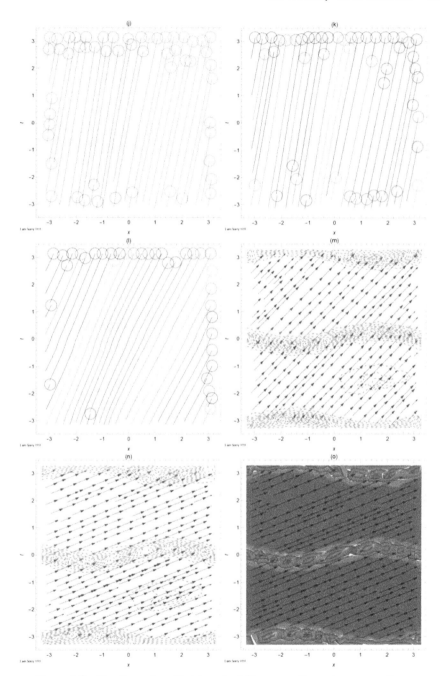

Figure 1.8 The dispersion effect in optical fibers regarding pulse propagation.

$$\mathscr{U}_{\mathrm{III},4}(x,t) = e^{i\int \frac{\left(\rho_2 \mathfrak{T}(x)^2 - \vartheta\right)^2}{4\rho_2(\varepsilon+\rho_1)\mathfrak{T}(x)^2} dx} e^{-it\vartheta}$$

$$\times \sqrt[4]{3} \sqrt{\frac{i\left(\gamma_2(\gamma_3-1)+4\gamma_3\sqrt{\gamma_2^2(\gamma_3-1)}\sinh^6\left(\frac{1}{2}\chi_2\right)\operatorname{csch}^4(\chi_2)\right)}{\sqrt{\gamma_3}\sqrt{\varsigma_2}\left(\gamma_3\tanh^2\left(\frac{1}{2}\chi_2\right)-1\right)}} \Big/ 2^{3/4}. \quad (1.22)$$

For $\gamma_1 \neq 0$, $\gamma_2 \neq 0$, and $\gamma_3 = \gamma_1\gamma_2$, we get

$$\mathscr{U}_{\mathrm{I},4}(x,t) = e^{i\int \frac{\left(\rho_2\mathfrak{T}(x)^2 - \vartheta\right)^2}{4\rho_2(\varepsilon+\rho_1)\mathfrak{T}(x)^2} dx} e^{-it\vartheta}$$

$$\times \sqrt{\frac{\left(\sqrt{a_0^2\left(9\gamma_2^2 - 32\gamma_1^2\gamma_2\right)} + 3a_0\gamma_2\right)\left(\gamma_1\sqrt{\frac{(4\gamma_1^2-\gamma_2)\gamma_2\tanh^2(\chi_3)\operatorname{sech}^2(\chi_3)}{\gamma_1^2}} - \gamma_2\operatorname{sech}^2(\chi_3)\right)}{2\gamma_2\left(\gamma_2 - 4\gamma_1^2\tanh^2(\chi_3)\right)}} + a_0, \quad (1.23)$$

$$\mathscr{U}_{\mathrm{II},5}(x,t) = e^{i\int \frac{\left(\rho_2 \mathfrak{T}(x)^2 - \vartheta\right)^2}{4\rho_2(\varepsilon+\rho_1)\mathfrak{T}(x)^2} dx} e^{-it\vartheta}$$

$$\times \sqrt{a_0 + \frac{i\sqrt{6}\sqrt{\gamma_1\gamma_2}\gamma_1\left(\gamma_1\sqrt{\frac{(4\gamma_1^2-\gamma_2)\gamma_2\tanh^2(\chi_3)\operatorname{sech}^2(\chi_3)}{\gamma_1^2}} - \gamma_2\operatorname{sech}^2(\chi_3)\right)}{\gamma_2\sqrt{\varsigma_2}\left(\gamma_2 - 4\gamma_1^2\tanh^2(\chi_3)\right)}}, \quad (1.24)$$

$$\mathscr{U}_{\mathrm{III},5}(x,t) = \frac{\sqrt[4]{3}}{2^{3/4}} e^{i\int \frac{\left(\rho_2\mathfrak{T}(x)^2 - \vartheta\right)^2}{4\rho_2(\varepsilon+\rho_1)\mathfrak{T}(x)^2} dx} e^{-it\vartheta}$$

$$\times \sqrt{\frac{i\left(\gamma_2^3 - 4\gamma_2^2\gamma_1^2\tanh^2(\chi_3) - 4\gamma_2^2\gamma_1^2\operatorname{sech}^2(\chi_3) + 4\gamma_2\gamma_1^3\sqrt{\frac{(4\gamma_1^2-\gamma_2)\gamma_2\tanh^2(\chi_3)\operatorname{sech}^2(\chi_3)}{\gamma_1^2}}\right)}{\gamma_2\sqrt{\gamma_1\gamma_2}\sqrt{\varsigma_2}\left(\gamma_2 - 4\gamma_1^2\tanh^2(\chi_3)\right)}}. \quad (1.25)$$

For $\gamma_2^2 - 4\gamma_1\gamma_3 = 0$, we get

$$\mathscr{U}_{1,5}(x,t) = e^{i\int \frac{\left(\rho_2 \mathfrak{T}(x)^2 - \vartheta\right)^2}{4\rho_2(\varepsilon+\rho_1)\mathfrak{T}(x)^2} dx} e^{-it\vartheta} \times \sqrt{\frac{\gamma_3\left(\sqrt{a_0^2\gamma_2^2} + 3a_0\gamma_2\right) e^{\gamma_2\varsigma}}{\gamma_2\left(e^{\frac{\gamma_2(x-\varepsilon t)}{2\sqrt{\gamma_3}}} - 2\gamma_3 e^{\gamma_2\varsigma}\right)}} + a_0, \quad (1.26)$$

$$\mathscr{U}_{\mathrm{II},6}(x,t) = e^{i\int \frac{\left(\rho_2 \mathfrak{T}(x)^2 - \vartheta\right)^2}{4\rho_2(\varepsilon+\rho_1)\mathfrak{T}(x)^2} dx} e^{-it\vartheta} \times \sqrt{a_0 + \frac{i\sqrt{\frac{3}{2}}\sqrt{\gamma_3}\gamma_2 e^{\gamma_2\varsigma}}{\sqrt{\varsigma_2}\left(e^{\frac{\gamma_2(x-\varepsilon t)}{2\sqrt{\gamma_3}}} - 2\gamma_3 e^{\gamma_2\varsigma}\right)}}, \quad (1.27)$$

$$\mathscr{U}_{\mathrm{III},6}(x,t) = e^{i\int \frac{\left(\rho_2 \mathfrak{T}(x)^2 - \vartheta\right)^2}{4\rho_2(\varepsilon+\rho_1)\mathfrak{T}(x)^2} dx} e^{-it\vartheta} \times \frac{\sqrt[4]{3}\sqrt{\frac{i\gamma_2}{\sqrt{\gamma_3}\sqrt{\varsigma_2}\left(1 - 2\gamma_3 e^{\gamma_2\varsigma - \frac{\gamma_2(x-\varepsilon t)}{2\sqrt{\gamma_3}}}\right)}}}{2^{3/4}}. \quad (1.28)$$

Here $\chi_1 = \sqrt{\gamma_1}(-\varepsilon t + x - \varsigma)$, $\chi_2 = \gamma_2(-\varepsilon t + x + 2\varsigma)$, and $\chi_3 = \sqrt{\gamma_1}(-\varepsilon t + x + \varsigma)$

1.3 Results and Discussion

In this section, we analyze the outcomes of the study as well as the computational strategies that were used. The $\mathcal{RR}$ equation has been solved using two different analytical approaches, including the Khater II method and the Sardar sub-equation method; several soliton wave solutions have been produced using a variety of different algorithms. The auxiliary equations of the employed techniques are comparable to one another. Eqn (1.6), (1.8), (1.10), (1.12), (1.13) and (1.14) that were used to obtain the results are depicted in the graphs referenced in the following references: Figures 1.1, 1.2, 1.3, 1.4, 1.5 and 1.6 when $\delta = -4$, $\varepsilon = 15$, $\varsigma_4 = 13$, $\varsigma_3 = 25$ & $\delta = -4$, $\varepsilon = -1$, $\varsigma_2 = -3$, $\varsigma_4 = -1$, $\varsigma_3 = -2$ & $\delta = -1$, $\varepsilon = 0$, $\varsigma_2 = 3$, $\varsigma_4 = 1$, $\varsigma_3 = 2$ & $a_0 = 6$, $\gamma_2 = 3$, $\gamma_1 = 2$, $\gamma_3 = 1$, $\varepsilon = 4$, $\varsigma = 7$ & $a_0 = 6$, $\gamma_2 = 3$, $\gamma_1 = 2$, $\gamma_3 = -1$, $\varepsilon = 4$, $\varsigma_2 = 4$ & $a_0 = 6$, $\gamma_2 = 3$, $\gamma_1 = 2$, $\gamma_3 = -1$, $\varepsilon = 4$, $\varsigma_2 = 4$, where each of these figures represents a different such as kink, periodic, soliton, and bright singular wave. In contrast, the interactions between the soliton wave solutions that were found are depicted by Figures 1.7 and 1.8, respectively. In conclusion, determining the originality of our findings requires comparing the solutions we created with those that have already been published in [1, 2, 3].

1.4 Conclusion

In this chapter, many soliton wave solutions have been constructed of the Rangwala–Rao equation in some distinct formulas such as rational, hyperbolic, and trigonometric. These solutions explain the effect of the integrability conditions on the electric field's slowly varying envelope. Besides, it impacts the slowly changing electric field envelope and the propagation of pulses in optical fibers regarding the dispersion effect. All these characterizations have been explained through some distinguished graphs. The paper's novelty and its scientific contributions have been discussed by comparing our results with some previously published articles. All solutions' accuracy has been checked by putting them back into the original model by Mathematica 13.1.

Declarations

Ethics approval and consent to participate
Not applicable.

Consent for publication
Not applicable.

Availability of data and material
The data that support the findings of this study are available from the corresponding author upon reasonable request.

Competing interests
The authors declare that they have no competing interests.

Funding
No fund has been received for this paper.

Authors' contributions
All authors contributed equally to the writing of this paper. All authors read and approved the final manuscript.

Acknowledgments
I greatly thank the journal staff (Editors and Reviewers) for their support and help.

References

[1] M. M. Khater, R. A. Attia, D. Lu, Modified auxiliary equation method versus three nonlinear fractional biological models in present explicit wave solutions, Mathematical and Computational Applications 24 (1) (2018) 1.

[2] M. M. Khater, M. S. Mohamed, R. A. Attia, On semi analytical and numerical simulations for a mathematical biological model; the time-fractional nonlinear kolmogorov–petrovskii–piskunov (kpp) equation, Chaos, Solitons & Fractals 144 (2021) 110676.

[3] M. Khater, C. Park, D. Lu, R. A. Attia, Analytical, semi-analytical, and numerical solutions for the cahn–allen equation, Advances in Difference Equations 2020 (1) (2020) 1–12.

[4] M. Khater, C. Park, D. Lu, R. A. Attia, Analytical, semi-analytical, and numerical solutions for the cahn–allen equation, Advances in Difference Equations 2020 (1) (2020) 1–12.

[5] M. M. Khater, D. Lu, R. A. Attia, Lump soliton wave solutions for the (2+ 1)-dimensional konopelchenko–dubrovsky equation and kdv equation, Modern Physics Letters B 33 (18) (2019) 1950199.

[6] C. Yue, M. Khater, R. A. Attia, D. Lu, The plethora of explicit solutions of the fractional ks equation through liquid–gas bubbles mix under the thermodynamic conditions via atangana–baleanu derivative operator, Advances in Difference Equations 2020 (1) (2020) 1–12.

[7] M. M. Khater, A. E.-S. Ahmed, M. El-Shorbagy, Abundant stable computational solutions of atangana–baleanu fractional nonlinear hiv-1 infection of cd4+ t-cells of immunodeficiency syndrome, Results in Physics 22 (2021) 103890.

[8] M. M. Khater, A. Mousa, M. El-Shorbagy, R. A. Attia, Analytical and semi-analytical solutions for phi-four equation through three recent schemes, Results in Physics 22 (2021) 103954.

[9] M. M. Khater, R. A. Attia, D. Lu, Explicit lump solitary wave of certain interesting (3+ 1)-dimensional waves in physics via some recent traveling wave methods, Entropy 21 (4) (2019) 397.

[10] M. M. Khater, Diverse solitary and jacobian solutions in a continually laminated fluid with respect to shear flows through the ostrovsky equation, Modern Physics Letters B 35 (13) (2021) 2150220.

[11] M. M. Khater, Abundant breather and semi-analytical investigation: On high-frequency waves' dynamics in the relaxation medium, Modern Physics Letters B 35 (22) (2021) 2150372.

[12] M. M. Khater, R. A. Attia, C. Park, D. Lu, On the numerical investigation of the interaction in plasma between (high & low) frequency of (langmuir & ion-acoustic) waves, Results in Physics 18 (2020) 103317.

[13] M. M. Khater, D. Lu, Analytical versus numerical solutions of the nonlinear fractional time–space telegraph equation, Modern Physics Letters B 35 (19) (2021) 2150324.

[14] M. M. Khater, Nonparaxial pulse propagation in a planar waveguide with kerr–like and quintic nonlinearities; computational simulations, Chaos, Solitons & Fractals 157 (2022) 111970.

[15] M. M. Khater, Lax representation and bi-hamiltonian structure of nonlinear qiao model, Modern Physics Letters B 36 (07) (2022) 2150614.

[16] M. M. Khater, Two-component plasma and electron trapping's influence on the potential of a solitary electrostatic wave with the dust-ion-acoustic speed, Journal of Ocean Engineering and Science (2022).

[17] M. M. Khater, D. Lu, Diverse soliton wave solutions of for the nonlinear potential kadomtsev–petviashvili and calogero–degasperis equations, Results in Physics 33 (2022) 105116.

[18] M. M. Khater, A. M. Alabdali, A. Mashat, S. A. Salama, et al., Optical soliton wave solutions of the fractional complex paraxial wave dynamical model along with kerr media, FRACTALS (fractals) 30 (05) (2022) 1–17.

[19] M. M. Khater, In solid physics equations, accurate and novel soliton wave structures for heating a single crystal of sodium fluoride, International Journal of Modern Physics B (2022) 2350068.

[20] M. M. Khater, De broglie waves and nuclear element interaction; abundant waves structures of the nonlinear fractional phi-four equation, Chaos, Solitons & Fractals 163 (2022) 112549.

[21] M. M. Khater, Novel computational simulation of the propagation of pulses in optical fibers regarding the dispersion effect, International Journal of Modern Physics B (2022) 2350083.

[22] M. M. Khater, Nonlinear elastic circular rod with lateral inertia and finite radius: Dynamical attributive of longitudinal oscillation, International Journal of Modern Physics B (2022) 2350052.

[23] M. Khater, Analytical and numerical-simulation studies on a combined mkdv–kdv system in the plasma and solid physics, The European Physical Journal Plus 137 (9) (2022) 1–9.

[24] M. Khater, Recent electronic communications; optical quasi–monochromatic soliton waves in fiber medium of the perturbed fokas–lenells equation, Optical and Quantum Electronics 54 (9) (2022) 1–12.

[25] M. M. Khater, Nonlinear biological population model; computational and numerical investigations, Chaos, Solitons & Fractals 162 (2022) 112388.

[26] D.-x. Kong, Explicit exact solutions for the lienard equation and its applications, Physics Letters A 196 (1-2) (1994) 301–306.

[27] X. Liu, Exact solitary wave solutions of the rangwala-rao equation, in: 2012 2nd International Conference on Uncertainty Reasoning and Knowledge Engineering, IEEE, 2012, pp. 175–178.

[28] Z. I. Al-Muhiameed, E. A.-B. Abdel-Salam, Generalized jacobi elliptic function solution to a class of nonlinear schrödinger-type equations, Mathematical Problems in Engineering 2011 (2011).

[29] Z. I. Al-Muhiameed, E. A.-B. Abdel-Salam, Generalized hyperbolic function solution to a class of nonlinear schrödinger-type equations, Journal of Applied Mathematics 2012 (2012).

[30] L. Cheng-Shi, The exact solutions to lienard equation with high-order nonlinear term and applications, Fizika A 18 (2009) 29.

2

Novel Cubic B-spline Based DQM for Studying Convection–Diffusion Type Equations in Extended Temporal Domains

J. P. Shukla[1], B. K. Singh[1], C. Cattani[2], and M. Gupta[1]

[1]School of Physical and Decision Sciences, Department of Mathematics,
Babasaheb Bhimrao Ambedkar University, India
[2]Engineering School, DEIM, University of Tuscia,
P.le dell'Università, Italy
E-mail: jaishukla1993@gmail.com; brijeshks@bbau.ac.in;
cattani@unitus.it; mukeshgupta.rs@bbau.ac.in

Abstract

This chapter aims to develop a new hybrid cubic B-spline differential quadrature method (in short nHCB-DQM) for studying convection–diffusion equations. The nHCB-DQM is based on DQM with new hybrid cubic B-splines as basis. The proposed nHCB-DQM together with temporal algorithm SSP-RK43 is utilized for finding numerical solutions of convection-diffusion equations in the extended temporal domain. The efficiency and effectiveness of the referred approach is examined via three different examples of convection–diffusion equation by means of absolute/L_2/L_∞ errors. The approximated results are compared with the exact results, and the existing results were published recently. Findings confirm that the evaluated results agree well with the exact solutions. Less data complexity, straightforwardness, and ease of implementation are the main advantages of nHCB-DQM.

Keywords: differential quadrature method, new hybrid cubic b-spline, convection–diffusion equation, SSP-RK43 formulae.

2.1 Introduction

The term convection in fluids is defined as the molecular movement and the diffusion as the particles spread via random motion starts from higher to lower concentration medium. The model of convection–diffusion equations is utilized to explain different kinds of chemical/physical phenomena. The numeric study of convection–diffusion is widely carried out in important models of engineering and science. A number of relevant examples such as fluid's transport through porous-medium or pollutant's transport through the atmosphere, the heat transfer through a permeable-medium, etc. Consider an initial-value system of nonlinear-type convection–diffusion equation (in short, C-D Eqn)

$$\begin{cases} \dfrac{\partial \omega}{\partial t} + \delta \dfrac{\partial \omega}{\partial x} - \alpha \dfrac{\partial^2 \omega}{\partial x^2} = N(\omega, \omega', \omega'') + G(x,t), \quad t > 0, \quad x \in [c,d] \\ \omega(x,0) = g(x), \end{cases} \quad (2.1)$$

with boundary conditions of Dirichlet's type:

$$\omega(c,t) = h_0(t), \quad \omega(d,t) = h_1(t) \quad (2.2)$$

where the parameter $\delta > 0$ is the phase speed and $\alpha > 0$ is the coefficient of viscosity. $g(x)$, $h_0(t)$, and $h_1(t)$ are the given sufficiently smooth functions. $N(\omega, \omega', \omega'')$ is selected as the potential energy and the nonlinear term. This problem may be seen in the computation of the modeling of fluid dynamics and hydraulics convection–diffusion of physical quantities such as heat, energy, mass, etc. [1].

The evaluation of the behavior of nonlinear partial differential equations, analytically, is a very complicated task, and so the numerical techniques play a very important role in approximating the behavior, which is close to the exact solution behavior. The researchers have developed numerous rigorous techniques for studying such equations. Some of the techniques are listed as collocation techniques [2–6], differential quadrature methods [7–15], finite difference scheme [16] and compact finite difference scheme [17], homotopy perturbation method [18], variation iteration method [19], J-transform-based optimal HAM and variational iteration (VIT) [20, 21], differential transform method [22–27], etc.

In the literature, numerous numerical schemes have been developed for computing 1D C-D Eqn with the coefficient of constant parameters [28–33]. Some latest schemes have presented numeric solutions to the C-D Eqn, such as: semi-discrete and a padé approximation method [34], fourth-order compact finite difference technique [35], CBCM with Neumann's BCs. [36], Chebyshev wavelets approximation of second kind [37], and comparative study of eqn (2.1) with three different techniques–in the first technique, the time derivative and space derivatives are respectively approximated by forward difference and PDQM, in the second technique, the time derivative and the space derivatives are approximated respectively by PDQM and central difference, while in the third technique, both derivatives (time and space) are approximated via PDQM [38], cubic quasi-interpolation spline-based collocation method [39], etc.

In this chapter, we have proposed DQM attached with so-called hybrid cubic B-splines (in short, nHCB-DQM) for spatial approximations of the derivatives in C-D Eqn (2.1), and, thus, the referred equation (2.1) transformed into a system of first-order ODEs in temporal space, which is enumerated via the well-known SSP-RK43 formulae [40]. The remainder of the chapter is organized as: Section 2.2 is a portrayal of nHCB-DQM, Section 2.3 explains the computation of $a_{il}^{(1)}$ and $a_{il}^{(2)}$, Section 2.4 elaborates the nHCB-DQM for the class of C-D Eqn (2.1), Section 2.5 presents numerical results and discussion, and concluding notes are presented in Section 2.6.

2.2 Portrayal of nHCB-DQM

Bellman and his associates have been given a computational method in 1972, which is called differential quadrature method (in short DQM) for describing the numeric solution and behavior of linear/nonlinear differential equations [41]. In this method, $\frac{\partial^k}{\partial x^r}$ of $\omega(x,t)$ at (x_i,t) for $i \in \Delta_n$ is approximated as follows [41, 42]:

$$\frac{\partial^k \omega(x_i,t)}{\partial x^k} = \sum_{\ell \in \Delta_n} a_{i\ell}^{(k)} \omega_\ell, \qquad i \in \Delta_n = \{1,2,\ldots,n\}, \; k=1,2., \qquad (2.3)$$

where ω_ℓ represents $\omega(x_\ell,t)$ and $a_{i\ell}^{(k)}$ be time-dependent unknown quantities, known as weighting coefficients (wt. coeff.) for the k th order derivative. These coefficient can be computed through a set of basis functions, since the wt. coeff. is based on grid spacing [42], and, therefore, a equi-spaced partition of the domain is taken, i.e., $\Lambda = \{x : c \leq x \leq d, c,d \in \mathbb{R}\}$ as $P[\Lambda] = \{x_i : h = x_{i+1} - x_i, \; x_i \in \Lambda, \; i \in \Delta_n\}$ with $c = x_1, x_{i-1} < x_i < x_{i+1}, x_n = d$

where $h = \frac{b-a}{n-1}$ is the discretization step in space. At the common grids, (x_i, t) denotes $\omega_i = \omega(x_i, t)$, $i \in \Delta_n$.

Now, the nHCB-splines function $nH_i^4 = nH_i^4(x)$ as in [43] is taken as

$$nH_i^4(x) = \begin{cases} -\frac{\mu}{6h^3}\mathbb{A}_{i-2}^3 + \frac{k_3(1-\mu)}{h^3\lambda}\{\lambda\mathbb{A}_{i-2} - \sinh(\lambda\mathbb{A}_{i-2})\}; & x \in [x_{i-2}, x_{i-1}) \\ \frac{\mu}{6h^3}(h^3 - 3h^2\mathbb{A}_{i-2} + 3h\mathbb{A}_{i-2}^2 + 3\mathbb{A}_{i-2}^3) + \\ \frac{(1-\mu)}{h^3}\{k_1 + k_2\mathbb{A}_i + k_4\exp(\lambda\mathbb{A}_i) + k_5\exp(\lambda\mathbb{A}_i)\}; & x \in [x_{i-1}, x_i) \\ \frac{\mu}{6h^3}(h^3 + 3h^2\mathbb{A}_{i+1} + 3h\mathbb{A}_{i+1}^2 - 3\mathbb{A}_{i+1}^3) + \\ \frac{(1-\mu)}{h^3}\{k_1 - k_2\mathbb{A}_i + k_4\exp(-\lambda\mathbb{A}_i) + k_5\exp(-\lambda\mathbb{A}_i)\}; & x \in [x_i, x_{i+1}) \\ \frac{\mu}{6h^3}\mathbb{A}_{i+2}^3 + \frac{k_3(1-\mu)}{h^3\lambda}\{-\lambda\mathbb{A}_{i+2} + \sinh(\lambda\mathbb{A}_{i+2})\}; & x \in [x_{i+1}, x_{i+2}) \\ 0; & \text{Else} \end{cases}$$

where

$$c = \cosh(\lambda h), \ s = \sinh(\lambda h), \ r = \frac{1}{(\lambda ch - s)(1-c)}, \ \mathbb{A}_i = (x_i - x);$$

$$k_1 = \lambda chr(1-c), \ k_2 = \frac{\lambda r}{2}\{s^2 - c(1-c)\}, \ k_3 = \frac{\lambda r(1-c)}{2},$$

$$k_4 = \frac{r}{4}\{\exp(-\lambda h)(-c+1) + s(-1+\exp(-\lambda h))\},$$

$$k_5 = \frac{r}{4}\{\exp(\lambda h)(-c+1) + s(-1+\exp(\lambda h))\}.$$

$\{nH_k^4 : k \in \Delta_{n+1} \cup \{0\}\}$ is utilized for the basis sets for the spatial derivatives over referred domain. The values of the base functions nH_i^4 and its derivatives of order 1 or 2 at common grids are depicted in Table 2.1, where

$$\kappa_1 = \frac{\mu}{6} + \frac{1-\mu}{2}\frac{s - ph}{phc - s}; \ \kappa_2 = \frac{2\mu}{3} + 1 - \mu;$$

Table 2.1 Values of $nH_i^4(x)$, $\frac{dnH_i^4}{dx}$, and $\frac{d^2 nH_i^4}{dx^2}$ at x_j

x	x_{j-2}	x_{j-1}	x_j	x_{j+1}	x_{j+2}
nH_i^4	0	κ_1	κ_2	κ_1	0
$\frac{dnH_i^4}{dx}$	0	κ_3	0	κ_4	0
$\frac{d^2 nH_i^4}{dx^2}$	0	κ_5	κ_6	κ_5	0

$$\kappa_3 = \frac{\mu}{2h} + \frac{1-\mu}{2}\frac{p(c-1)}{phc-s} = -\kappa_4;$$

$$\kappa_5 = \frac{\mu}{h^2} + \frac{1-\mu}{2}\frac{p^2 s}{pch-s} = -\frac{\kappa_6}{2}.$$

Analogous to the modification of base functions given in [45], nHCB-spline base functions are modified as follows:

$$\begin{cases} \theta_1(*) = nH_1^4(*) + 2nH_0^4(*) \\ \theta_2(*) = nH_2^4(*) - nH_1^4(*) \\ \vdots \\ \theta_i(*) = nH_i^4(*), \ i \in \Delta_{n-1}\setminus\{1\} \\ \vdots \\ \theta_{n-1}(*) = nH_{n-1}^4(*) - nH_{n+1}^4(*) \\ \theta_n(*) = nH_n^4(*) + 2nH_{n+1}^4(*) \end{cases}.$$

Thus, $\{\theta_k : k \in \Delta_n\}$ represents the modified base functions of the derivatives over the respective domain.

2.3 Computation of Wt. Coeff. $a_{il}^{(1)}$ and $a_{il}^{(2)}$

To find the wt. coeff. $a_{il}^{(k)}$, nHCB-DQM is considered in modified form as the set of base functions. Write $\theta_{pi} = \theta_p(x_i)$, $\theta'_{pi} = \theta'_p(x_i)$, and $\theta'''_{pi} = \theta'''_p(x_i)$. The approximate values of the first-order derivatives in DQM are obtained by the following relations:

$$\theta'_{pi} = \sum_{l \in \Delta_n} a_{il}^{(1)} \theta_{pl}, \quad p, i \in \Delta_n. \tag{2.4}$$

Now setting the wt. coeff. matrixes $A = \left[a_{il}^{(1)}\right]$, $\Theta = [\theta_{il}]$, and $\Theta' = [\theta'_{il}]$, eqn (2.3) can be recast into a compact matrix form as follows:

$$\Theta A^T = \Theta', \tag{2.5}$$

where the coefficient matrix

$$\Theta = \begin{bmatrix} \kappa_2+2\kappa_1 & \kappa_1 & & & & & \\ 0 & \kappa_2 & \kappa_1 & & & & \\ & \kappa_1 & \kappa_2 & \kappa_1 & & & \\ & & \ddots & \ddots & \ddots & & \\ & & & \kappa_1 & \kappa_2 & \kappa_1 & \\ & & & & \kappa_1 & \kappa_2 & 0 \\ & & & & & \kappa_1 & \kappa_2+2\kappa_1 \end{bmatrix}.$$

In matrix Θ', the columns are read as follows:
$\Theta'[1] = [2\kappa_4, 2\kappa_3, 0, \ldots, 0, 0]^T$, $\Theta'[2] = [\kappa_4, 0, \kappa_3, 0, \ldots, 0, 0]^T, \ldots, \Theta'[n-1] = [0, 0, \ldots, \kappa_4, 0, \kappa_3]^T$, and $\Theta'[n] = [0, \ldots, 0, 2\kappa_4, 2\kappa_3]^T$. Keep in mind that the nHCB-spline functions are modified in such a way that the coeff. matrix Θ in eqn (2.5) is diagonally dominant. Thereafter, for computing the wt. coeff. $a_{il}^{(1)}$ from the system (2.5), we adopt the well-known Thomas algorithm.

The basis functions of any finite-dimensional vector space are infinitely many. Therefore, it gives many opportunities to compute the wt. coeff. in the same domain by different basis functions. The computation of wt. coeff. $a_{il}^{(2)}$ is established on advantages of polynomial DQM. Once wt. coeff. $a_{il}^{(1)}$ of first-order derivatives are computed, second-order derivatives of wt. coeff. $a_{il}^{(2)}$ can be computed via a recursive formula [42] given as follows:

$$\begin{cases} a_{il}^{(k)} = k\left(a_{il}^{(1)} a_{ii}^{(k-1)} - \dfrac{a_{il}^{(k-1)}}{x_i - x_l}\right), & i, l \in \Delta_n;\ i \neq l \\ a_{il}^{(r)} = -\sum\limits_{i \in \Delta_n \setminus \{l\}} a_{il}^{(r)}, & l \in \Delta_n;\ i = l \end{cases} \quad (2.6)$$

2.4 The nHCB-DQM for the Class of C–D Eqn

The initial value system (2.1) is re-written in the following form:

$$\frac{\partial \omega}{\partial t} = \alpha \frac{\partial^2 \omega}{\partial x^2} - \delta \frac{\partial \omega}{\partial x} + N(\omega, \omega', \omega'') + G(x.t). \quad (2.7)$$

After using nHCB-DQM and boundary conditions (2.2), we have

$$\frac{\partial \omega_i}{\partial t} = \alpha \sum_{\ell \in \Delta_{n-1} \setminus \{1\}} a_{i\ell}^{(2)} \omega_\ell - \delta \sum_{\ell \in \Delta_{n-1} \setminus \{1\}} a_{i\ell}^{(1)} \omega_\ell +$$
$$N\left(\omega_i, \sum_{\ell \in \Delta_{n-1} \setminus \{1\}} a_{i\ell}^{(1)} \omega_\ell, \sum_{\ell \in \Delta_{n-1} \setminus \{1\}} a_{i\ell}^{(2)} \omega_\ell\right) + H_i, \quad (2.8)$$

where $H_i = \alpha \sum_{\ell \in \{1,n\}} a_{i\ell}^{(2)} \omega_\ell - \delta \sum_{\ell \in \{1,n\}} a_{i\ell}^{(1)} \omega_\ell + F_i + G_i$, $G_i = G(x_i, t)$ and F_i is the boundary part of the nonlinear term.

Now, the above reduced system of ODEs (2.8) in time has been solved via strongly stable scheme (SSP-RK43 scheme) [40] as follows:

$$\wp^{(1)} = \omega^m + \frac{\Delta t}{2} L(\omega^m), \wp^{(2)} = \wp^{(1)} + \frac{\Delta t}{2} L(\wp^{(1)}),$$
$$\wp^{(3)} = \frac{2}{3} \omega^m + \frac{\wp^{(2)}}{3} + \frac{\Delta t}{6} L(\wp^{(2)}); \omega^{m+1} = \wp^{(3)} + \frac{\Delta t}{2} L(\wp^{(3)}).$$

The evaluation now prefers to start from eqn (2.8) via SSP-RK43, which needs an initial solution, and this can be taken from $\omega_i(t=0) = g_i$ for $i = 2, 3, \ldots, n-1$ (initial conditions given with the C-D Eqn).

2.5 Numerical Results and Discussion

In this section, three examples of C-D Eqn are considered to check the performance of nHCB-DQM by calculating absolute L_2 and L_∞ error norms for measuring the efficiency, consistency, and accuracy of the scheme. L_2 and L_∞ error norms are defined as follows:

$$L_\infty := \max_{k \in \Delta_n} \{|\omega_k - \omega_k^*|\} \quad \text{and} \quad L_2 := \sqrt{h \sum_{k=1}^{n} |\omega_k - \omega_k^*|}, \quad (2.9)$$

where ω_k^* = exact solution; ω_k = enumerated solution at x_k.

Spatial order of convergence of the utilized approach is computed using

$$\frac{\ln(E_r(n_1)) - \ln(E_r(n_2))}{\ln(n_2) - \ln(n_1)},$$

where $E_r(n_i)$ denotes the prescribed error with grid-points n_i ($i = 1, 2$).

Table 2.2 The absolute L_2 and L_∞ errors at $h = 0.01$ for example 2.1.

	Present method			RCBCM [45]			Douglas Method		
x	$t=1$	$t=2$	$t=5$	$t=1$	$t=2$	$t=5$	$t=1$	$t=2$	$t=5$
0.1	4.54E-07	6.25E-07	6.11E-07	1.73E-07	2.29E-07	2.58E-07	1.33E-04	1.77E-04	2.00E-04
0.5	9.66E-07	1.23E-06	1.13E-06	5.24E-07	9.13E-07	1.36E-06	4.04E-04	7.02E-04	1.05E-03
0.9	2.00E-08	4.89E-08	1.95E-08	5.37E-07	8.09E-07	1.12E-06	4.15E-04	6.30E-04	8.83E-04
L_2	8.02E-07	1.01E-06	9.71E-07						
L_∞	1.58E-06	1.01E-06	1.78E-06						
μ	3.000001	3.000001	3.000001						

Example 2.1 *In the first example, consider the C-D Eqn (2.1)*

$$\frac{\partial \omega}{\partial t} + \delta \frac{\partial \omega}{\partial x} - \alpha \frac{\partial^2 \omega}{\partial x^2} = 0, \quad t > 0, \quad 0 \leq x \leq 1.$$

As per requirement, the initial/boundary conditions can be evaluated from the exact solution $\omega(x,t) = \exp(c_1 x + c_2 t)$. For the numeric computation, parameters are taken as $c_1 = 1.17712434446770$, $c_2 = -0.09$, $\alpha = 0.02$, and $\delta = 0.1$ [45].

In the Figure 2.1, the absolute errors and nHCB-DQ at three different time slots $t = 1, 2, 5$ are depicted for fixed parameter values $c_1 = 1.17712434446770$, $c_2 = -0.09$, $\alpha = 0.02$, and $\delta = 0.1$ at $\mu = 3.000001$

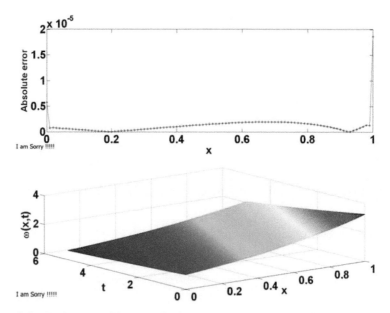

Figure 2.1 At the top and bottom, absolute errors ($t = 5$) and physical behaviour ($t \leq 5$) plots of nHCB-DQM solutions, respectively, for Example 2.1.

(optimal value). These findings are tested with the existing results RCBCM [45] and Douglas method reported in Table 2.2.

The absolute error/physical behavior of C-D Eqn (2.1) is depicted in Figure 2.1 for $t \leq 5$. Findings of the present method shows that the computed outcomes (with optimal μ) are improved upon the existing results due to RCBCM [45] and the Douglas method.

Example 2.2 In the second example, consider the C-D Eqn (2.1)

$$\frac{\partial \omega}{\partial t} + \delta \frac{\partial \omega}{\partial x} - \alpha \frac{\partial^2 \omega}{\partial x^2} = 0, \quad t > 0, \quad 0 \leq x \leq 1.$$

As per requirement, the initial/boundary conditions can be evaluated from the exact solution $\omega(x,t) = \exp(c_1 x + c_2 t)$. For the numerical computation, we take $c_1 = 0.02854797991928$, $c_2 = -0.0999$, $\alpha = 0.022$, and $\delta = 3.5$ [45].

In Figure 2.2, the absolute errors and solution computed via nHCB-DQMs at three different time slots $t = 1, 2, 5$ with the fixed values of parameter, $c_1 = 0.02854797991928$, $c_2 = -0.0999$, $\alpha = 0.022$, and $\delta = 3.5$ and the optimal values of $\mu = 3.000038, 3.000038,$ and 3.000037 are

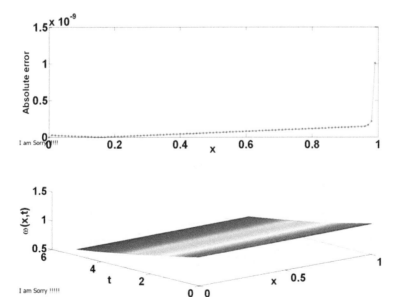

Figure 2.2 At the top and bottom, absolute errors ($t = 5$) and physical behavior ($t \leq 5$) plots of nHCB-DQM solutions, respectively, for Example 2.2.

Table 2.3 The absolute L_2 and L_∞ errors at $h = 0.01$ for Example 2.2.

x	Present method			[33]			Douglas method		
	t=1	t=2	t=5	t=1	t=2	t=5	t=1	t=2	t=5
0.1	1.70E-10	1.54E-10	9.79E-12	2.56E-10	2.38E-10	5.65E-10	2.56E-07	2.37E-07	5.63E-07
0.5	7.00E-11	6.33E-11	6.09E-11	8.39E-10	1.38E-09	1.91E-09	8.37E-07	1.38E-06	1.90E-06
0.9	3.23E-11	2.92E-11	1.33E-10	1.33E-09	2.83E-09	3.97E-09	1.33E-06	2.82E-06	3.95E-07
L_2	1.78E-10	1.61E-10	1.29E-10						
L_∞	1.47E-09	1.33E-09	1.01E-09						
μ	3.000038	3.000038	3.000037						

depicted. These findings are tested with the existing results [33] and the Douglas method, reported in Table 2.3. Figure 2.2 depicts the absolute and physical behaviour of C-D Eqn (2.1) for $t \leq 5$. Findings of the present method shows that the computed outcomes (with optimal μ) are comparatively more accurate than existing results due to [33] and the Douglas method.

Example 2.3 *In the third example, consider the nonlinear C-D Eqn (2.1)*

$$\frac{\partial \omega}{\partial t} + \delta \frac{\partial \omega}{\partial x} - \alpha \frac{\partial^2 \omega}{\partial x^2} = \omega \frac{\partial \omega}{\partial x} - 2x(\omega - 1) - 1, \quad t > 0, \quad 0 \leq x \leq 3.$$

As per requirement, the initial/boundary conditions can be evaluated from the exact solution $\omega(x,t) = x^2 + t$. For the numeric computation, parameters are taken as $\alpha = 1.0$ and $\delta = 1.0$ [46].

Figure 2.3 depicts absolute errors and nHCB-DQ solution at three different time slots $t = 1, 3, 5$ for fixed parameter values: $\alpha = 1.0$, $\delta = 1.0$, and

Table 2.4 Absolute error for Example 2.3, with $\mu = 3.5$ and $h = 0.016667$ at $t = 1, 3, 5$.

	Present method		
x	t=1	t=3	t=5
0.25	5.24E-05	5.48E-05	5.53E-05
0.5	5.10E-05	5.52E-05	5.58E-05
0.75	5.03E-05	5.55E-05	5.60E-05
1	5.05E-05	5.57E-05	5.61E-05
1.25	5.16E-05	5.57E-05	5.61E-05
1.5	5.29E-05	5.58E-05	5.61E-05
1.75	5.42E-05	5.58E-05	5.62E-05
2	5.53E-05	5.58E-05	5.62E-05
2.25	5.53E-05	5.58E-05	5.62E-05
2.5	5.55E-05	5.58E-05	5.62E-05
2.75	5.55E-05	5.58E-05	5.62E-05
L_2	9.21E-05	9.61E-05	9.68E-05
L_∞	6.04E-05	6.07E-05	6.11E-05

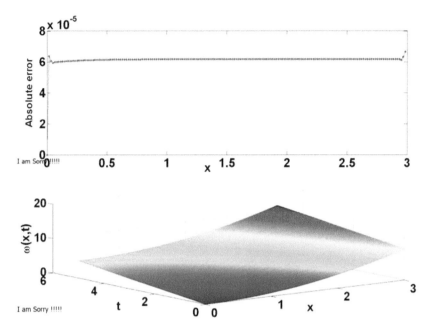

Figure 2.3 At the top and bottom, absolute errors ($t = 5$) and physical behaviour ($t \leq 5$) plots of nHCB-DQM solutions, respectively, for Example 2.3.

$\mu = 3.5$, reported in Tab. 2.4. Figure 2.3 depicts the absolute error (physical behavior) of C-D Eqn (2.1) for $t \leq 5$. The outcomes confirm that nHCB-DQ results with optimal μ report less variability in errors.

2.6 Conclusion

In this work, we have introduced a new type of base functions with differential quadrature method named as nHCB-DQM to give the numeric solution of C-D Eqn (2.1). In nHCB-DQM, base functions are used for the first time in modified form to compute the first-order wt. coeff. Thereafter, Shu's recursive formula is used to compute the second-order wt. coeff. After implementing the newly introduced nHCB-DQM with the required initial/boundary conditions, C-D Eqn (2.1) is converted into the system of ODEs in time with initial conditions. Followed by a well-known scheme, SSP-RK43 [40] is used to solve the final system of ODEs and get the required results. The findings show that the computed results are improved upon the recent methods: RCBCM [45], [33], and the Douglas method.

References

[1] PJ Roache. Computational fluid dynamics. *Computational Fluid Dynamics*, 1976.

[2] RC Mittal and Geeta Arora. Numerical solution of the coupled viscous burgers' equation. *Commu. Nonli. Sci. Numer. Simul.*, 16(3):1304–1313, 2011.

[3] S. Singh, S. Singh, and R Arora. Numerical solution of second-order one-dimensional hyperbolic equation by exponential b-spline collocation method. *Numer. Anal. Appli.*, 10(2):164–176, 2017.

[4] G Arora, RC Mittal, and BK Singh. Numerical solution of bbm-burger equation with quartic b-spline collocation method. *J. Eng. Sci. Tech.*, 9: 104–116, 2014.

[5] M Ramezani, H. Jafari, SJ Johnston, and D. Baleanu. Complex b-spline collocation method for solving weakly singular volterra integral equations of the second kind. *Miskolc Mathematical Notes*, 16(2): 1091–1103, 2015.

[6] M. Abbas, Ahmad Abd Majid, Ahmad Izani Md. Ismail, and A. Rashid. Numerical method using cubic b-spline for a strongly coupled reaction-diffusion system. *PLoS One*, 9(1):e83265, 2014.

[7] G Arora and BK Singh. Numerical solution of burgers' equation with modified cubic b-spline differential quadrature method. *Appl. Math. Comput.*, 224:166–177, 2013.

[8] BK Singh and P. Kumar. A novel approach for numerical computation of burgers' equation in (1+ 1) and (2+ 1) dimensions. *Alex. Eng. J.*, 55 (4):3331–3344, 2016.

[9] BK Singh. A novel approach for numeric study of 2d biological population model. *Cogent Mathematics*, 3(1):1261527, 2016.

[10] BK Singh and G. Arora. A numerical scheme to solve fisher-type reaction-diffusion equations. *Nonlinear Studies/Mesa-Mathematics in Engineering, Science and Aerospace*, 5(2):153–164, 2014.

[11] BK Singh, G Arora, and MK Singh. A numerical scheme for the generalized burgers–huxley equation. *Journal of the Egyptian Mathematical Society*, 24(4):629–637, 2016.

[12] BK Singh and C Bianca. A new numerical approach for the solutions of partial differential equations in three-dimensional space. *Appl. Math. Inf. Sci*, 10(5):1–10, 2016.

[13] B.K. Singh, J.P. Shukla, and M. Gupta. Study of one-dimensional hyperbolic telegraph equation via a hybrid cubic b-spline differential quadrature method. *Int. J. Appl. Comput. Math.*, 7(1), 2021.

[14] BK Singh and P. Kumar. An algorithm based on a new dqm with modified extended cubic b-splines for numerical study of two dimensional hyperbolic telegraph equation. *Alex. Eng. J*, 57(1):175–191, 2018.

[15] BK Singh and P. Kumar. An algorithm based on dqm with modified trigonometric cubic b-splines for solving coupled viscous burgers equations. *arXiv preprint arXiv:1611.04792*, 2016.

[16] VK Srivastava and BK Singh. A robust finite difference scheme for the numerical solutions of two dimensional time dependent coupled nonlinear burgers' equations. *Int. J. of Appl. Math and Mech*, 10(7): 28–39, 2014.

[17] BK Singh, G. Arora, and P. Kumar. A note on solving the fourth-order kuramoto-sivashinsky equation by the compact finite difference scheme. *Ain Shams Eng. J.*, 9(4):1581–1589, 2018.

[18] BK Singh, P. Kumar, and V. Kumar. Homotopy perturbation method for solving time fractional coupled viscous burgers' equation in (2+1) and (3+1) dimensions. *Int. J. Appl. Comput. Math.*, 4:38, 2018.

[19] BK Singh and P. Kumar. Fractional variational iteration method for solving fractional partial differential equations with proportional delay. *Int. J. Differ. Eqn*, 2017:1–11, 2017. URL https://doi.org/10.1155/2017/5206380.

[20] BK Singh, A. Kumar, and M. Gupta. Efficient new approximations for space-time fractional multi-dimensional telegraph equation. *Int. J. Appl. Comput. Math*, 8:218, 2022.

[21] BK Singh and A. Kumar. New approximate series solutions of conformable time–space fractional fokker–planck equation via two efficacious techniques. *Partial Differential Equations in Applied Mathematics*, 6(100451), .

[22] BK Singh and P. Kumar. Extended fractional reduced differential transform for solving fractional partial differential equations with proportional delay. *Int. J. Appl. Comput. Math.*, 3(1), .

[23] BK Singh and S. Agrawal. A new approximation of conformable time fractional partial differential equations with proportional delay. *Appl. Numer. Math.*, 157, .

[24] BK Singh and M. Gupta. A comparative study of analytical solutions of space-time fractional hyperbolic like equations with two reliable methods. *Arab. J. Basic Appl. Sci.*, 26(1).

[25] V.K. Singh, BK andSrivastava. Approximate series solution of multi-dimensional, time fractional-order (heat-like) diffusion equations using frdtm. *R. Soc. Open Sci.*, 2(5).
[26] BK Singh and P. Kumar. Frdtm for numerical simulation of multi-dimensional, time-fractional model of navier-stokes equation. *Ain Shams Eng. J.*, 9(4), .
[27] BK Singh and S. Agrawal. Study of time fractional proportional delayed multi-pantograph system and integro-differential equations. *Math. Meth. Appli. Sci.*, 45, .
[28] M. Dehghan. Weighted finite difference techniques for the one-dimensional advection–diffusion equation. *Appl. Math. Comput.*, 147(2):307–319, 2004.
[29] M. Sari, G. Gürarslan, and A. Zeytinoğlu. High-order finite difference schemes for solving the advection-diffusion equation. *Math. Comput. Appl.*, 15(3):449–460, 2010.
[30] A. Mohebbi and M. Dehghan. High-order compact solution of the one-dimensional heat and advection–diffusion equations. *Appl. Math. Modelling*, 34(10):3071–3084, 2010.
[31] DK Salkuyeh. On the finite difference approximation to the convection–diffusion equation. *Appl. Mathematics Comput.*, 179(1):79–86, 2006.
[32] H. Karahan. Implicit finite difference techniques for the advection–diffusion equation using spreadsheets. *Adv. Engg. Soft.*, 37(9):601–608, 2006.
[33] Hassan NA Ismail, Elsayed ME Elbarbary, and Ghada SE Salem. Restrictive taylor's approximation for solving convection–diffusion equation. *Appl. Math. Comput.*, 147(2):355–363, 2004.
[34] H. Ding and Y. Zhang. A new difference scheme with high accuracy and absolute stability for solving convection–diffusion equations. *J. Comput. Appl. Math.*, 230(2):600–606, 2009.
[35] H-H. Cao, L.-B. Liu, Y. Zhang, and S.-M. Fu. A fourth-order method of the convection–diffusion equations with neumann boundary conditions. *Appl. Math. Comput.*, 217(22):9133–9141, 2011.
[36] RC Mittal and RK Jain. Numerical solution of convection-diffusion equation using cubic b-splines collocation methods with neumann's boundary conditions. *Int. J. Appl. Math. Comput.*, 4(2):115–127, 2012.
[37] F. Zhou and X. Xu. Numerical solution of the convection diffusion equations by the second kind chebyshev wavelets. *Appl. Math. Comput.*, 247:353–367, 2014.

[38] VS Aswin, A. Awasthi, and C Anu. A comparative study of numerical schemes for convection-diffusion equation. *Procedia Engineering*, 127: 621–627, 2015.

[39] S Bouhiri, A. Lamnii, and M Lamnii. Cubic quasi-interpolation spline collocation method for solving convection–diffusion equations. *Math. Comput. Simul.*, 164:33–45, 2019.

[40] RJ Spiteri and SJ Ruuth. A new class of optimal high-order strong-stability-preserving time discretization methods. *SIAM J. Numeri. Anal.*, 40(2):469–491, 2002.

[41] R. Bellman, BG Kashef, and J Casti. Differential quadrature: a technique for the rapid solution of nonlinear partial differential equations. *J. Comput. Phy.*, 10(1):40–52, 1972.

[42] C. Shu and BE Richards. Application of generalized differential quadrature to solve two-dimensional incompressible navier-stokes equations. *Int. J. Numer. Methods in Fluids*, 15(7):791–798, 1992.

[43] I. Wasim, M. Abbas, and M. Amin. Hybrid b-spline collocation method for solving the generalized burgers-fisher and burgers-huxley equations. *Math. Prob. Engg*, 2018, 2018.

[44] RC Mittal and R. Bhatia. Numerical solution of second order one dimensional hyperbolic telegraph equation by cubic b-spline collocation method. *Appl. Math. Comput.*, 220:496–506, 2013.

[45] RC Mittal and RK Jain. Redefined cubic b-splines collocation method for solving convection–diffusion equations. *Appl. Math. Modelling*, 36 (11):5555–5573, 2012.

[46] M. Abolhasani, S. Abbasbandy, and T. Allahviranloo. A new variational iteration method for a class of fractional convection-diffusion equations in large domains. *Mathematics*, 5(2):26, 2017.

3

Study of the Ranking-function-based Fuzzy Linear Fractional Programming Problem: Numerical Approaches

Ravinder Kaur[1], Rakesh Kumar[1], and Hatıra Günerhan[2]

[1]Lovely Professional University, India
[2]Department of Mathematics, Faculty of Education, Kafkas University, Turkey
E-mail: ravinder.kaur@lpu.co.in; rakeshmalhan23@gmail.com; gunerhanhatira@gmail.com

Abstract

This chapter has highlighted the study of the fuzzy linear fractional programming problem that considered coefficients of objective function and constraints as trapezoidal fuzzy number. To obtain an optimal solution, utilization of ranking functions has been employed to trapezoidal fuzzy number to convert the fuzzy FLPP to crisp FLPP, and for further computation, the crisp fractional LPP is solved with the Charan Cooper method to transform fractional programming into LPP and the same crisp FLPP using complementary LPP. An illustration from real life has been used to demonstrate the effectiveness of our suggested strategy.

Keywords: ranking function, numerical techniques, symmetric trapezoidal number.

3.1 Introduction

The fractional programming problem, also known as the FPP, is a problem that requires careful decision-making in situations where one is attempting to maximize an existing ratio that is restricted. Linear fractional programming is a mathematical technique for optimally allocating resources to diverse tasks based on predetermined criteria of optimality. LFP is a very important tool in the field of operations research. Further methods for applying it to real-world problems have been developed within the framework of the LFP problem. Applications of linear fractional objectives can be found in the financial sector, the hospitality industry, military academies, and the conversion of foreign loans into local ones. Researchers have proposed fuzzy optimization for situations where there is ambiguity in the available options. In order to deal with incorrect input in the LFP problem of set membership, we have introduced the more advanced level of interactive fuzzy programming and applied multiple choice fuzzy set theories.

Tanaka et al. [1] created fuzzy programming based on Bellman and Zadeh's fuzzy decision framework [2], and the linear programming problem was the first problem they used fuzzy settings to solve. Lai and Hwang [3], Shaocheng [4], and others thought about what happens when all the parameters are unclear. Researchers used this concept frequently to address FLFP challenges [5, 6, 7]. Dehghan et al. [8] proposed a fuzzy linear programming approach for precisely resolving fuzzy linear programming problems. Buckley et al. [10, 11–13] presented a method to approximate the solution to completely fuzzy linear programming problems.

In this study, we discussed a way to find the fuzzy optimum FLFP solution that fits within the limits of an inequality. The restrictions on the left and right sides are shown as real numbers, while the coefficients of the goal function are shown as symmetric trapezoidal fuzzy numbers. This chapter presents a form of fuzzy arithmetic for symmetric trapezoidal fuzzy numbers that can be applied to FLFP problems.

3.2 Preliminaries

This section is to cover the basic concepts to fuzzy sets theory and trapezoidal fuzzy numbers.

Definition 2.1 [5]. A fuzzy set $\widetilde{\mathcal{F}}$ with membership functions $\mu_{\widetilde{\mathcal{F}}}(x) : \mathbb{R} \to [0, 1]$ represents a fuzzy number if the following characteristics are met:

i) $\mu_{\widetilde{\mathcal{F}}}(\lambda x + (1 - \lambda) y) \geq \min \{\mu_{\widetilde{\mathcal{F}}}(x), \mu_{\widetilde{\mathcal{F}}}(y)\}$, for every $x, y \in \mathcal{X}$ and $\lambda \in [0, 1]$, i.e., $\widetilde{\mathcal{F}}$ is convex

ii) $\widetilde{\mathcal{F}}$ is normal, i.e., $\mu_{\widetilde{\mathcal{F}}}(x_0) = 1$.

iii) $\mu_{\widetilde{\mathcal{F}}}(x)$ is piecewise continuous.

Definition 2.2 [5]. Fuzzy number $\tilde{\mathcal{A}}=(a,b,c,d)$ is said to be a trapezoidal fuzzy number when $\mu_{\tilde{\mathcal{A}}}(x)$, its membership function, is defined as

$$\mu_{\tilde{\mathcal{A}}}(x) = \begin{cases} \frac{(x-a)}{(b-a)}, & \text{when } a \leq x \leq b \\ 1 & \text{when } b \leq x \leq c \\ \frac{(d-x)}{(d-c)}, & \text{when } c \leq x \leq d \\ 0 & \text{otherwise} \end{cases}.$$

Definition 2.3 [5]. For two symmetric trapezoidal fuzzy integers' arithmetic operations
Let $\tilde{\mathcal{A}} = (a_l, a_U, \alpha, \alpha)$ and $\tilde{\mathcal{B}} = (b_l, b_U, \beta, \beta)$, then

$$\tilde{\mathcal{A}} + \tilde{\mathcal{B}} = (a_l + b_l, a_U + b_U, \alpha + \beta, \alpha + \beta),$$
$$\tilde{\mathcal{A}} - \tilde{\mathcal{B}} = (a_l - b_l, a_U - b_U, \alpha - \beta, \alpha - \beta),$$

$$\tilde{\mathcal{A}}\tilde{\mathcal{B}} = \left(\left(\frac{a_l + a_U}{2}\right)\left(\frac{b_l + b_U}{2}\right) - t, \left(\frac{a_l + a_U}{2}\right)\left(\frac{b_l + b_U}{2}\right) + t, |a_U \beta + b_U \alpha|, |a_U \beta + b_U \alpha|\right),$$

where

$$t = \frac{t_2 - t_1}{2}, \quad t_1 = \min\{a_l b_l, a_U b_U, a_U b_l, a_l b_U\},$$
$$t_1 = \max\{a_l b_l, a_U b_U, a_U b_l, a_l b_U\},$$
$$k\tilde{\mathcal{A}} = \begin{cases} (ka_l, ka_U, k\alpha, k\alpha), & k \geq 0 \\ (ka_l, ka_U, -k\alpha, -k\alpha), & k < 0 \end{cases}.$$

Definition 2.4 [5]. Let $F(\mathbb{R})$ represent fuzzy numbers set and then the function $\mathcal{R}$ from $F(\mathbb{R})$ to $\mathbb{R}$ defines the ranking functions for trapezoidal fuzzy numbers of kind $\tilde{\mathcal{A}} = (a,b,c,d)$ with $\mathcal{R}_1(\tilde{\mathcal{A}}) = \frac{a+2b+2c+d}{6}$.

Definition 2.5 [5]. Let $F(\mathbb{R})$ represent fuzzy numbers set and then the function $\mathcal{R}$ from $F(\mathbb{R})$ to $\mathbb{R}$ defines the ranking functions for symmetric trapezoidal fuzzy numbers $\tilde{\mathcal{A}} = (a_l, a_U, \alpha, \alpha)$ $\mathcal{R}(\tilde{\mathcal{A}}) = \frac{(a_l + a_U)}{2}$.

3.3 General Form of Fuzzy LFPP

The general model of LFPP is expressed as follows:

$$\max Z(x) = \frac{p^T x + \alpha}{q^T x + \beta} = \frac{\mathcal{M}(x)}{\mathcal{N}(x)}, \qquad (3.1)$$

subject to $x \in \mathcal{S} = \{x \in \mathbb{R}^n : \mathcal{A}x. \leq \mathcal{B}, x \geq 0\}$ with $=\mathcal{A} \in \mathbb{R}^{m \times n}$, $\mathcal{B}. \in \mathbb{R}^n$, p^T, $q^T \in \mathbb{R}^n$ and α, $\beta \in \mathbb{R}$.

To avoid those values of x for which $\mathcal{N}(x)$ is zero, it is required that for the system $\mathcal{A}x. \leq \mathcal{B}$ with $x \geq 0$, either $\mathcal{N}(x) > 0$ or $\mathcal{N}(x) < 0$.

For the sake of convenience, we assume that LFP satisfies the conditions keeping $\mathcal{N}(x) > 0$.

Remark: If $\mathcal{M}(x)$ is concave on $\mathcal{S}$ with $\mathcal{M}(x)$ convex and positive on $\mathcal{S}$, then problem (3.1) represents standard concave–convex programming problem.

(i) By Charnes and Coopers [1] method using linear transformation $y = \tau x$ LFPP (3.1) can be reduced to equivalent LPP as follows:

$$\max \tau \, \mathcal{M}(y/\tau), \qquad (3.2)$$

subject to

$$\mathcal{A}(y/\tau) - \mathcal{B} \leq 0,$$

$$\tau \mathcal{N}(y/\tau) \leq 1,$$

$$y \geq 0 \text{ and } \tau > 0.$$

If there are more than one linear fractional objective function known as multi-objective linear fractional programming problem whose general form is expressed as follows:

$$\max Z_i(x) = \frac{p_i^T x + \alpha_i}{q_i^T x + \beta_i} = \frac{\mathcal{M}_i(x)}{\mathcal{N}_i(x)}, i = 1, 2, \ldots, k,$$

subject to $x \in \mathcal{S} = \{x \in \mathbb{R}^n : \mathcal{A}x. \leq \mathcal{B}, x \geq 0\}$ with $\mathcal{A}. \in \mathbb{R}^{m \times n}$, $\mathcal{B}. \in \mathbb{R}^n$, p_i^T, $q_i^T \in \mathbb{R}^n$ and α_i, $\beta_i \in \mathbb{R}$.

(ii) Solution of FLPP with complementary method.

By this approach, the FLPP model expressed in eqn (3.1) can be reduced to its equivalent LPP as follows:

$$\mathrm{Max} Z^* = Z_1 - Z_2,$$

where
$$Z_1 = \max \mathcal{M}(x) \quad \text{and} \quad Z_2 = \min \mathcal{N}(x),$$
with $\mathcal{A}x. \leq \mathcal{B}$ with $x \geq 0$.

3.4 Algorithm for the Solution of FLFPP with Trapezoidal Fuzzy Number TrpFN

In this section, we will show how to solve a FLFPP using the following procedures.

Step 1: Consider the LFPP in which the objective function and constraints have trapezoidal fuzzy parameters.

Step 2: Using ranking functions given above in preliminaries section under Definition 2.4 or 2.5, convert an FFLPP to a crisp FLPP.

Step 3: Convert FLPP to crisp LPP with Charan and Cooper transformation method to find optimal solution.

Step 4: Convert FLPP to crisp LPP with a complementary method to find the optimal solution.

Step 5: Compare the results obtained from the two methods to draw conclusions.

3.5 Numerical Example

In this section, an illustration is given to apply the above-discussed algorithm using real-life situations.

In a city, a baker produces two kinds of energy bars: one nut-based energy bar N and another cereal-based energy bar C. On N, it gets profit around (4,6,3,3) and on C around (1,5,1,1) rupees per unit, respectively. (3,7,2,2) and (3,1,1,1) rupees are the costs of N and C, respectively. It is assumed that a fixed cost of approximately (1,1,2,2) is added to the cost function due to the anticipated duration throughout the production process. Let us say that the raw materials needed to make product A and product B cost about 3 and 5 units per kilogram, respectively, and that the supply of these raw materials is limited to about 15 kilograms. Each manufactured unit of N requires about 5 man-hours of labor, while each manufactured unit of C requires about 2 man-hours of labor. However, each day, only about 10 man-hours of labor are available. Find the optimal production volume of N and C energy bars to maximize profits.

The above problem can be expressed as the following LPP:

$$\text{Max } \mathcal{W} = \text{Max} \frac{(4,6,3,3)\,u_1 + (1,5,1,1)u_2}{(3,7,2,2)\,u_1 + (3,1,1,1)\,u_2 + (1,1,2,2)}, \quad (3.3)$$

subject to

$$3u_1 + 5u_2 \leq 15,$$
$$5u_1 + 2u_2 \leq 10,$$
$$u_1, u_2 \geq 0.$$

Using rank function for symmetric trapezoidal fuzzy number $\widetilde{A} = (a_l,\ a_U,\ \alpha, \alpha)$ and $\mathcal{R}\left(\widetilde{A}\right) = \frac{(a_l + a_U)}{2}$ will reduce as

$$\text{Max}\mathcal{W} = \text{Max} \frac{5u_1 + 3u_2}{5u_1 + 2u_2 + 1}, \quad (3.4)$$

subject to

$$3u_1 + 5u_2 \leq 15,$$
$$5u_1 + 2u_2 \leq 10,$$
$$u_1, u_2 \geq 0.$$

Now the above crisp LFPP is converted to crisp LPP by two different techniques to get the optimal solution.

Method 1:

Let $v_1 = \tau u_1$ and $v_2 = \tau u_2$; then the crisp LFP problem (3.4) can be transformed to LPP using the Charnes and Coopers method as follows:

$$Max\ \mathcal{W} = 5v_1 + 3v_2,$$

subject to

$$3v_1 + 5v_2 - 15\tau \leq 0,$$
$$5v_1 + 2v_2 - 10\tau \leq 0,$$
$$5v_1 + 2v_2 + \tau \leq 1,$$
$$v_1, v_2, \tau \geq 0.$$

Then the optimal solution obtained at $u_1 = 0$ and $u_2 = 2.9993$ is $\mathcal{W} = 1.2857$.

Method 2:
Convert the LFPP to LPP with complementary technique:

$$Max\ \mathcal{W}_1 = 5u_1 + 3u_2,$$

subject to
$$3u_1 + 5u_2 \leq 15,$$
$$5u_1 + 2u_2 \leq 10,$$
$$u_1, u_2 \geq 0,$$

and
$$Min\ \mathcal{W}_1 = 5u_1 + 2u_2 + 1,$$

subject to
$$3u_1 + 5u_2 \leq 15,$$
$$5u_1 + 2u_2 \leq 10,$$
$$u_1, u_2 \geq 0,$$
$$Max\ \mathcal{W}^* = u_2 - 1,$$

subject to
$$3u_1 + 5u_2 \leq 15,$$
$$5u_1 + 2u_2 \leq 10,$$
$$u_1, u_2 \geq 0.$$

Using the simplex method, we get the optimal solution obtained at $u_1 = 0, u_2 = 3$ as $\mathcal{W} = 1.29$.

3.6 Conclusion

The attention in this chapter is given to find the solution of fuzzy LFPP with trapezoidal fuzzy number. Two different methods were discussed to find the optimal solution. The approaches discussed to yield the optimal solution are to deal with the complexity in problem solving by converting into the equivalent model, which is easy to compute.

Acknowledgements

The authors oblige and value the anonymous reviewer's insightful criticism and recommendations.

References

[1] H. Tanaka, T. Okuda, K. Asai, 'Fuzzy mathematical programming', Transactions of the society of instrument and control engineers, 9(5), pp. 607-613. 1973

[2] R. E.Bellman, L. A.Zadeh, 'Decision-making in a fuzzy environment', Management science, 17(4), B- 141, 1970

[3] Y. J.Lai, C. L. Hwang, 'A new approach to some possibilistic linear programming problems', Fuzzy sets and systems, 49(2), pp. 121-133,1992.

[4] T. Shaocheng, 'Interval number and fuzzy number linear programming', Fuzzy sets and systems, 66(3), pp. 301-306, 1994.

[5] A. Ebrahimnejad, & M. Tavana, 'A novel method for solving linear programming problems with symmetric trapezoidal fuzzy numbers', Applied mathematical modelling, 38(17-18), pp. 4388-4395, 2014.

[6] F. H.Lotfi, T.Allahviranloo, M. A.Jondabeh, L. Alizadeh, 'Solving a full fuzzy linear programming using lexicography method and fuzzy approximate solution', Applied mathematical modelling, 33(7), pp.3151-3156, 2009.

[7] H. S. Najafi, S. A. Edalatpanah, 'A note on a new method for solving fully fuzzy linear programming problems', Applied mathematical modelling, 37(14-15), pp.7865-7867, 2013.

[8] M. Dehghan, B. Hashemi, M. Ghatee, 'Computational methods for solving fully fuzzy linear systems', Applied mathematics and computation, 179(1), pp.328-343, 2006.

[9] A. Charnes and W. W. Cooper, Programming with linear fractional functions, Nav. Res. Log. Quar. 9 , pp. 181-186, 1962.

[10] G. R. Bitran, A. G. Novaes, Linear programming with a fractional objective function, Oper. Res.21, 1973, pp. 22-291.

[11] L. A. Zadeh, Fuzzy Sets, Inf. and Cont. 8, pp. 338-353, 1965.

[12] R. E. Bellman, L. A. Zadeh, Decision making in a fuzzy environment, Manag. Sci. Vol. 17, pp. 141-164, 1970.

[13] H. J. Zimmermann, Description and optimization of fuzzy systems, Internat. J. General Systems Vol.2, 1 pp. 209-215, 1976

4

Orthogonal Collocation Approach for Solving Astrophysics Equations using Bessel Polynomials

Shelly Arora and Indu Bala

Department of Mathematics, Panjabi University, India
E-mail: aroshelly@gmail.com; indu13121994@gmail.com

Abstract

A convenient computational algorithm combining Newton–Raphson method and orthogonal collocation using Bessel polynomials $J_n(\xi)$ as base function has been proposed to solve the Lane–Emden type equations. Numerical examples have been discussed to check the reliability and efficiency of the scheme. Numerically calculated results have been compared to the exact values as well as the values already given in the literature to check the compatibility of the scheme. Error analysis has been discussed by calculating the absolute errors, L_2 norm and L_∞ norm. Computer codes have been prepared using MATLAB.

Keywords: astrophysics equations, hypergeometric functions, bessel function, legendre polynomials, polytropic fluid

4.1 Introduction

The law of nature and physical world are commonly described in the form of differential equations. These equations play a significant role in physics, engineering, biology, economics etc. Therefore, a number of computational methods have been developed to obtain the solution of these differential equations [14]. A number of mathematicians and scientists showed their interest toward second-order differential equations. One such example is the Lane–Emden equation named after two astrophysicists Jonathan Lane and Robert Emden. The Lane–Emden equation has many phenomena in physics, quantum mechanics, and astrophysics [2, 13, 18]. The Lane–Emden equation is a singular linear as well as nonlinear boundary value problem that is a dimensionless form of Poisson's equation [2]. Due to the singularity behavior at the origin, the Lane–Emden equation becomes a challenging problem.

In the present chapter, the orthogonal collocation method with Bessel polynomials as base functions has been followed to numerically discretize the Lane–Emden equation. This method is called the Bessel collocation method (BCM). It was first introduced by Yuzbasi [2, 18] to solve a system of differential equations. In this method, a trial function is introduced with Bessel polynomials $J_i(\xi)$, where $i = 1, 2, 3, ..., n$ and $\xi \in (a, b)$, as base functions [5, 10, 16, 21]. This trial function is fitted to both the differential equations and boundary conditions and the residual is set equal to zero at collocation points.

Collocation method [15] is useful to find the numerical solution of functional equations. The numerical solution $\theta(\xi)$ is obtained by using the trial function that satisfies the functional equation $L_v(\theta) = 0$ at collocation points [20]. The collocation points are taken to be the roots of orthogonal polynomials. Zeros of Legendre polynomials have been taken as collocation points.

The general form of the Lane–Emdon type equations is represented as follows:

$$\frac{d^2\theta(\xi)}{d\xi^2} + \frac{a}{\xi}\left(\frac{d\theta(\xi)}{d\xi}\right) + g(\xi, \theta(\xi)) = 0. \tag{4.1}$$

The second-order nonlinear ordinary Lane–Emden differential equation [2, 3] for the polytropic fluid in spherical symmetry when $a = 2$ and $g(\theta, \xi) = \theta^m(\xi)$ is

$$\frac{1}{\xi^2}\frac{d}{d\xi}\left(\xi^2 \frac{d\theta(\xi)}{d\xi}\right) + \theta^m(\xi) = 0, \tag{4.2}$$

where m is a real constant, ξ is a dimensionless radius, and θ is related to the density. In the polytropic fluid, the index m is defined as a polytropic index. By differentiating eqn(4.2) with respect to ξ

$$\frac{d^2\theta(\xi)}{d\xi^2} + \frac{2}{\xi}\left(\frac{d\theta(\xi)}{d\xi}\right) + \theta^m(\xi) = 0. \tag{4.3}$$

4.2 Bessel Collocation Method

The general form of the Lane–Emden equation of first and second kinds is

$$\frac{d^2\theta(\xi)}{d\xi^2} + \frac{2}{\xi}\left(\frac{d\theta(\xi)}{d\xi}\right) + f(\xi)\theta^p(\xi) = 0 \qquad (4.4)$$

$$\frac{d^2\theta(\xi)}{d\xi^2} + \frac{2}{\xi}\left(\frac{d\theta(\xi)}{d\xi}\right) + g(\xi)e^{q\theta} = 0, \qquad (4.5)$$

where p and q are taken to be real constants and the functions $f(\xi)$ and $g(\xi)$ are arbitrary real-valued functions. If isothermal fluids are used instead of taking polytropic fluids, then the Lane–Emden equation reduces to the Emden–Chandershekhar equation [13]. An American astrophysicist, Subrmanyam Chandersekhar, has introduced the Chandershekhar white dwarf equation. It was introduced on the basis of gravitational potential of completely degenerate white dwarf. The Chandershekhar white dwarf equation is

$$\frac{1}{\xi^2}\frac{d}{d\xi}\left(\xi^2\frac{d\theta(\xi)}{d\xi}\right) + (\theta^2 - C)^{\frac{3}{2}} = 0, \qquad (4.6)$$

where θ represents the density of the white dwarf and C is any real constant and is related to the density of the white dwarf at the center. For $C = 0$, eqn (4.6) reduces to the Lane–Emden equation with $m = 3$.

4.2 Bessel Collocation Method

Orthogonal collocation is one of the weighted residual methods to solve linear and nonlinear differential equations. In this method, a trial function is proposed to adjust the differential equation along with boundary conditions [15, 20]. Orthogonal collocation includes interior, boundary, and mixed collocation depending upon the trial function adjusted in differential equation or boundary conditions or both, accordingly [7, 6, 14, 20]. In this method, zeros of an orthogonal polynomial are taken to be collocation points from where the term orthogonal collocation evolves. The residual is set orthogonal to weight functions of the given polynomial. In the proposed method, Bessel polynomials $J_n(\xi)$ of order n have been used as base functions to discretize the second-order linear or nonlinear differential equation. After applying the collocation principle, the linear as well as nonlinear differential equation converts into algebraic equation. At different collocation points, there are different algebraic equations and these equations are solved by iterative techniques such as the Newton–Raphson method [8, 11]. To find the numerical solution of the following model problem:

$$\theta''(\xi) = f(\xi, \theta(\xi), \theta'(\xi)); a \le \xi \le b, \qquad (4.7)$$

with general boundary conditions

$$a_1 \theta(a) - a_2 \theta'(a) = d_1; a_i \ge 0, i = 1, 2$$

$$b_1 \theta(b) + b_2 \theta'(b) = d_2; b_i \ge 0, i = 1, 2,$$

where d_1 and d_2 are constants and $f(\xi, \theta(\xi), \theta'(\xi))$ may be linear or nonlinear functional equation of ξ, θ, and θ'.

In the proposed method, numerical approximation of second-order differential equation is expressed in terms of Bessel series as follows:

$$\theta(\xi) = \sum_{i=1}^{n+1} c_i J_i(\xi); a \le \xi \le b, \qquad (4.8)$$

where c_i's are unknown constants and $\theta(\xi)$ is considered as an approximate numerical solution of eqn (4.7). In eqn (4.8), $i = 1, 2, 3, \ldots n+1$, and n is defined to be a positive integer such that $n \ge 1$. Using Bessel polynomials, the numerical approximation of differential equation of the proposed problem is obtained. In hypergeometric form as given in [1, 16], the Bessel function can be written as follows:

$$J_n(\xi) = \frac{\xi^n}{2^n n!} {}_0F_1(-; n+1; -\frac{1}{4}\xi^2). \qquad (4.9)$$

The first-order derivative of the Bessel function is defined as follows:

$$\frac{d}{d\xi}(\xi^n J_n(\xi)) = \xi^n J_{n-1}(\xi) \qquad (4.10)$$

$$\frac{d}{d\xi}(\xi^{-n} J_n(\xi)) = -\xi^{-n} J_{n+1}(\xi). \qquad (4.11)$$

The Bessel coefficients also follow from the power series expansion, which shows that for small values of ξ [11]

$$\lim_{\xi \to 0} {}_0F_1(-; n+1; -\frac{1}{4}\xi^2) = 1.$$

$$\lim_{\xi \to 0} \xi^{-n} J_n(\xi) = \frac{1}{2^n n!},$$

4.2 Bessel Collocation Method

which shows that the Bessel coefficient $J_n(\xi)$ at small values of ξ approaches $\frac{\xi^n}{2^n n!}$.

In the present chapter, the Bessel collocation method has been applied on single configuration $\theta(\xi)$. In this method, a trial function $\theta(\xi)$ has been approximated in terms of Bessel polynomials as follows:

$$\theta(\xi) = \sum_{i=1}^{n+1} J_i(\xi) c_i(t), \qquad (4.12)$$

where $J_i(\xi)$ are i^{th} order Bessel polynomials. Bessel polynomials can be written as suggested by [2, 19, 18, 17] to simplify eqn (4.12).

$$\theta(\xi) = \sum_{i=1}^{n+1} \xi^{i-1} K c_i(t), \qquad (4.13)$$

where K is a square matrix of order $(n+1) \times (n+1)$.

For n being an odd integer, K is defined as follows:

$$K = \begin{bmatrix} \frac{1}{0!0!2^0} & 0 & 0 & \cdots & 0 & 0 \\ 0 & \frac{1}{0!1!2^1} & 0 & \cdots & 0 & 0 \\ \frac{-1}{1!1!2^2} & 0 & \frac{1}{0!2!2^2} & \cdots & 0 & 0 \\ \vdots & \vdots & \vdots & \vdots & \vdots & \vdots \\ \vdots & \vdots & \vdots & \vdots & \vdots & \vdots \\ \frac{(-1)^{\frac{n-1}{2}}}{(\frac{n-1}{2})!(\frac{n-1}{2})!2^{n-1}} & 0 & \frac{(-1)^{\frac{n-3}{2}}}{(\frac{n-3}{2})!(\frac{n+1}{2})!2^{n-1}} & \cdots & \frac{1}{0!(n-1)!2^{n-1}} & 0 \\ 0 & \frac{(-1)^{\frac{n-1}{2}}}{(\frac{n-1}{2})!(\frac{n+1}{2})!2^n} & 0 & \cdots & 0 & \frac{1}{0!n!2^n} \end{bmatrix}$$

However, for n being an even integer, the matrix K can be written as follows:

$$K = \begin{bmatrix} \frac{1}{0!0!2^0} & 0 & 0 & \cdots & 0 & 0 \\ 0 & \frac{1}{0!1!2^1} & 0 & \cdots & 0 & 0 \\ \frac{-1}{1!1!2^2} & 0 & \frac{1}{0!2!2^2} & \cdots & 0 & 0 \\ \vdots & \vdots & \vdots & \ddots & \vdots & \vdots \\ \vdots & \vdots & \vdots & \ddots & \vdots & \vdots \\ 0 & \frac{(-1)^{\frac{n-2}{2}}}{(\frac{n-2}{2})!(\frac{n}{2})!2^{n-1}} & 0 & \cdots & \frac{1}{0!(n-1)!2^{n-1}} & 0 \\ \frac{(-1)^{\frac{n}{2}}}{(\frac{n}{2})!(\frac{n}{2})!2^n} & 0 & \frac{(-1)^{\frac{n-2}{2}}}{(\frac{n-2}{2})!(\frac{n+2}{2})!2^n} & \cdots & 0 & \frac{1}{0!n!2^n} \end{bmatrix}$$

To simplify eqn (4.13) at the jth collocation point:

$$\theta(\xi_j, t) = \sum_{i=1}^{n+1} \xi_j^{i-1} K c_i(t), \qquad j = 1, 2, ..., n+1. \tag{4.14}$$

Now, rewrite eqn (4.14) in matrix form at the jth collocation point

$$[\theta_j] = [X][\mathbf{c}], \tag{4.15}$$

where $X = [\xi_j^{i-1}]K$ and θ_j represents the value of θ at the jth collocation point and $\mathbf{c}$ represents the matrix of unknown collocation functions $c_i(t)$.

$$[X]^{-1}[\theta_j] = [\mathbf{c}]. \tag{4.16}$$

Substituting collocation coefficients from eqn (4.16) to eqn(4.13):

$$\theta(\xi) = \sum_{i=1}^{n+1} \xi^{i-1} X^{-1} \theta_i. \tag{4.17}$$

Again after simplifying, eqn (4.17) can be written in terms of Lagrangian interpolation polynomials [15] as follows:

$$\theta(\xi) = \sum_{i=1}^{n+1} l_i(\xi) \theta_i. \tag{4.18}$$

The Lagrangian interpolation polynomial at j^{th} collocation point can be represented by $l_i(\xi)$ and is calculated as follows:

$$l_i(\xi) = \Psi(\xi)/[(\xi - \xi_i)\Psi'(\xi_i)], \tag{4.19}$$

where $\Psi(\xi) = \xi(1-\xi)\prod_{j=1}^{n-1}(\xi - \xi_j)$. Details of Lagrangian interpolation polynomial have been discussed in [15]. Lagrangian interpolation polynomials help to obtain the numerical solutions over the desired node points by using the solution of $\theta(\xi)$ found by the Bessel collocation method over collocation points. The discretized form of first- and second-order derivatives of the collocation functions $\theta(\xi)$ at the jth collocation point can be calculated as follows:

$$\Psi^k(\xi_j) = \Psi'(\xi_i)[l_i^{(k)}(\xi_j)(\xi_j - \xi_i) + k l_i^{(k-1)}(\xi_j)]. \tag{4.20}$$

4.2 Bessel Collocation Method

First- and second-order derivative of $l_i(\xi)$ for $i = j$ are obtained by substituting $k = 2$ and 3, respectively, in eqn (4.20):

$$l'_i(\xi_j) = \Psi''(\xi_j)/2\Psi'(\xi_j)$$

and

$$l''_i(\xi_j) = \Psi'''(\xi_j)/3\Psi'(\xi_j).$$

First- and second-order derivative of $l_i(\xi)$ for $i \neq j$ are obtained by substituting $k = 1$ and 2, respectively, in eqn (4.20):

$$l'_i(\xi_j) = \frac{\Psi'(\xi_j)}{\Psi'(\xi_i)(\xi_j - \xi_i)}$$

and

$$l''_i(\xi_j) = \frac{\Psi''(\xi_j)}{\Psi'(\xi_i)(\xi_j - \xi_i)} - \frac{2l'_i(\xi_j)}{(\xi_j - \xi_i)}.$$

Being polynomial function, the higher order derivatives of $\Psi(\xi)$ exist at ξ_j and can be computed as follows:

$$\Psi'(\xi) = \sum_{m_1=1}^{n+1} \frac{1}{(\xi - \xi_{m_1})} \prod_{i=1}^{n+1} (\xi - \xi_i)$$

$$\Psi''(\xi) = 2! \sum_{m_2=1, m_2 \neq m_1}^{n+1} \frac{1}{(\xi - \xi_{m_2})} \left[\sum_{m_1=1}^{n+1} \frac{1}{(\xi - \xi_{m_1})} \prod_{i=1}^{n+1} (\xi - \xi_i) \right]$$

$$\Psi'''(\xi) = 3! \sum_{m_3=1, m_3 \neq m_2, m_1}^{n+1} \frac{1}{(\xi - \xi_{m_3})} \times$$

$$\left[\sum_{m_2=1, m_2 \neq m_1}^{n+1} \frac{1}{(\xi - \xi_{m_2})} \left[\sum_{m_1=1}^{n+1} \frac{1}{(\xi - \xi_{m_1})} \prod_{i=1}^{n+1} (\xi - \xi_i) \right] \right].$$

First- and second-order derivatives of eqn (4.18) are given as follows:

$$\theta'_j(\xi) = \sum_{i=1}^{n+1} A_{ji} \theta_i \qquad (4.21)$$

and

$$\theta''_j(\xi) = \sum_{i=1}^{n+1} B_{ji} \theta_i. \qquad (4.22)$$

By substituting eqn (4.18), (4.21), and (4.22) in eqn (4.17), following system algebraic equations are formed:

$$\sum_{i=1}^{n+1} B_{ji}\theta_i = f\left(\xi_j, \theta_j, \sum_{i=1}^{n+1} A_{ji}\theta_i\right), \qquad a \leq \xi \leq b;.$$

Similarly, the general equation of the Lane–Emden type equation (4.1) can be discretized to the system of algebraic equations by using eqn (4.18)–(4.22) as follows:

$$\sum_{i=1}^{n+1} B_{ji}\theta_i + \frac{a}{\xi_j}\sum_{i=1}^{n+1} A_{ji}\theta_i + g(\xi_j, \theta_j) = 0. \qquad (4.23)$$

The terms A_{ji} and B_{ji} are first- and second-order discretized forms of derivatives of $l_i(x)$ at the jth collocation point, respectively. Boundary conditions for $\theta(x,t)$ configuration are assumed to be $\theta(a,t) = \theta_1 = \theta_a$ and $\theta(b,t) = \theta_{n+1} = \theta_b$.

Block matrix structure of the generalized form of eqn (4.23) is represented as follows:

$$B_\theta + \mathcal{X}A_\theta + G = 0. \qquad (4.24)$$

In the matrix representation of eqn (4.24), there is no effect of boundary conditions as they are in scalar form and merge into G.

$$\mathcal{X} = \begin{bmatrix} \frac{a}{\xi_2} \\ \frac{a}{\xi_3} \\ \frac{a}{\xi_4} \\ \vdots \\ \frac{a}{\xi_n} \end{bmatrix}_{(n-1)\times 1} \qquad G = \begin{bmatrix} g(\xi_2, \theta_2) \\ g(\xi_3, \theta_3) \\ g(\xi_4, \theta_4) \\ \vdots \\ g(\xi_n, \theta_n) \end{bmatrix}_{(n-1)\times 1}$$

$$B_\theta = B.\Theta \qquad A_\theta = A.\Theta$$

$$\Theta = \begin{bmatrix} \theta_2 \\ \theta_3 \\ \theta_4 \\ \vdots \\ \theta_n \end{bmatrix}_{(n-1)\times 1}$$

4.2 Bessel Collocation Method

$$A = \begin{bmatrix} A(2,2) & A(2,3) & A(2,4) & A(2,5) & \cdots & \cdots & A(2,n) \\ A(3,2) & A(3,3) & A(3,4) & A(3,5) & \cdots & \cdots & A(3,n) \\ A(4,2) & A(4,3) & A(4,4) & A(4,5) & \cdots & \cdots & A(4,n) \\ \vdots & \vdots & \vdots & \vdots & \ddots & \ddots & \vdots \\ \vdots & \vdots & \vdots & \vdots & \ddots & \ddots & \vdots \\ A(n,2) & A(n,3) & A(n,4) & A(n,5) & \cdots & \cdots & A(n,n) \end{bmatrix}_{(n-1)\times(n-1)}$$

$$B = \begin{bmatrix} B(2,2) & B(2,3) & B(2,4) & B(2,5) & \cdots & \cdots & B(2,n) \\ B(3,2) & B(3,3) & B(3,4) & B(3,5) & \cdots & \cdots & B(3,n) \\ B(4,2) & B(4,3) & B(4,4) & B(4,5) & \cdots & \cdots & B(4,n) \\ \vdots & \vdots & \vdots & \vdots & \ddots & \ddots & \vdots \\ \vdots & \vdots & \vdots & \vdots & \ddots & \ddots & \vdots \\ B(n,2) & B(n,3) & B(n,4) & B(n,5) & \cdots & \cdots & B(n,n) \end{bmatrix}_{(n-1)\times(n-1)}.$$

The matrix representation corresponds to a system of (n-1) equations forming a block matrix structure given in eqn (4.24) and can be solved numerically by using iterative methods. To solve the system of algebraic equations, the Newton–Raphson method has been followed. The system of algebraic equations defined by eqn (4.24) can be generalized as follows:

$$\mathcal{F}_1(\theta_2(\xi), \theta_3(\xi), \theta_4(\xi), ..., \theta_n(\xi)) = 0$$

$$\mathcal{F}_2(\theta_2(\xi), \theta_3(\xi), \theta_4(\xi), ..., \theta_n(\xi)) = 0$$

$$\mathcal{F}_3(\theta_2(\xi), \theta_3(\xi), \theta_4(\xi), ..., \theta_n(\xi)) = 0$$

$$\vdots$$

$$\mathcal{F}_{n-1}(\theta_2(\xi), \theta_3(\xi), \theta_4(\xi), ..., \theta_n(\xi)) = 0.$$

These $n-1$ algebraic equations depend on $n-1$ variables $\theta_2(\xi)$, $\theta_3(\xi)$, $\theta_4(\xi)$, ..., $\theta_n(\xi)$. These variables represent $\theta(\xi_2)$, $\theta(\xi_3)$, $\theta(\xi_4)$, ..., $\theta(\xi_n)$, respectively. As the derivatives exist for all algebraic equations, the Newton–Raphson method can be followed to discretize the system of equations. For the solution to these algebraic equations, a function can be defined as follows:

$$M = \begin{bmatrix} \mathcal{F}_1(\theta_2(\xi), \theta_3(\xi), \theta_4(\xi), ..., \theta_n(\xi)) \\ \mathcal{F}_2(\theta_2(\xi), \theta_3(\xi), \theta_4(\xi), ..., \theta_n(\xi)) \\ \mathcal{F}_3(\theta_2(\xi), \theta_3(\xi), \theta_4(\xi), ..., \theta_n(\xi)) \\ \vdots \\ \mathcal{F}_{n-1}(\theta_2(\xi), \theta_3(\xi), \theta_4(\xi), ..., \theta_n(\xi)) \end{bmatrix}. \quad (4.25)$$

Now, the Newton–Raphson method for the system of algebraic equations can be applied by choosing the initial value approximation θ_2^0, θ_3^0, $\theta_4^0,......,\theta_n^0$ for $\theta_2, \theta_3, \theta_4, ... ,\theta_n$, respectively. Afterwards, the Jacobian matrix is computed to solve the system of equations iteratively.

Let $\mathcal{Q}^0$ be the matrix of all initial approximations θ_2^0, θ_3^0, $\theta_4^0,......,\theta_n^0$ such as

$$\mathcal{Q}^0 = \begin{bmatrix} \theta_2^0(\xi) \\ \theta_3^0(\xi) \\ \theta_4^0(\xi) \\ \vdots \\ \theta_n^0(\xi) \end{bmatrix}. \tag{4.26}$$

According to the procedure of the Newton–Raphson method, put the assumed initial approximations in the matrix of system of eqn (4.25) such as follows:

$$M^0 = \begin{bmatrix} \mathcal{F}_1(\theta_2^0(\xi), \theta_3^0(\xi), \theta_4^0(\xi), ..., \theta_n^0(\xi)) \\ \mathcal{F}_2(\theta_2^0(\xi), \theta_3^0(\xi), \theta_4^0(\xi), ..., \theta_n^0(\xi)) \\ \mathcal{F}_3(\theta_2^0(\xi), \theta_3^0(\xi), \theta_4^0(\xi), ..., \theta_n^0(\xi)) \\ \vdots \\ \mathcal{F}_{n-1}(\theta_2^0(\xi), \theta_3^0(\xi), \theta_4^0(\xi), ..., \theta_n^0(\xi)) \end{bmatrix} \tag{4.27}$$

Now, the obtained Jacobian ($\mathcal{J}$) is

$$\mathcal{J} = \mathcal{J}(\theta_2, \theta_3, ..., \theta_n)$$

$$\mathcal{J} = \begin{bmatrix} \frac{\partial \mathcal{F}_1}{\partial \theta_2(\xi)} & \frac{\partial \mathcal{F}_1}{\partial \theta_3(\xi)} & \frac{\partial \mathcal{F}_1}{\partial \theta_4(\xi)} & \cdots & \frac{\partial \mathcal{F}_1}{\partial \theta_n(\xi)} \\ \frac{\partial \mathcal{F}_2}{\partial \theta_2(\xi)} & \frac{\partial \mathcal{F}_2}{\partial \theta_3(\xi)} & \frac{\partial \mathcal{F}_2}{\partial \theta_4(\xi)} & \cdots & \frac{\partial \mathcal{F}_2}{\partial \theta_n(\xi)} \\ \frac{\partial \mathcal{F}_3}{\partial \theta_2(\xi)} & \frac{\partial \mathcal{F}_3}{\partial \theta_3(\xi)} & \frac{\partial \mathcal{F}_3}{\partial \theta_4(\xi)} & \cdots & \frac{\partial \mathcal{F}_3}{\partial \theta_n(\xi)} \\ \vdots & \vdots & \vdots & \vdots & \vdots \\ \frac{\partial \mathcal{F}_{n-1}}{\partial \theta_2(\xi)} & \frac{\partial \mathcal{F}_{n-1}}{\partial \theta_3(\xi)} & \frac{\partial \mathcal{F}_{n-1}}{\partial \theta_4(\xi)} & \cdots & \frac{\partial \mathcal{F}_{n-1}}{\partial \theta_n(\xi)} \end{bmatrix}, \tag{4.28}$$

where $\mathcal{J}$ is an $(n-1) \times (n-1)$ matrix. Let $\mathcal{J}^0$ be the matrix obtained by substituting initial approximation θ_2^0, θ_3^0, ... ,θ_n^0 in above Jacobian matrix equation (4.28) such as

$$\mathcal{J}^0 = \mathcal{J}(\theta_2^0, \theta_3^0, ..., \theta_n^0).$$

Let

$$h^0 = -\left[\mathcal{J}^0\right]^{-1}_{(n-1)\times(n-1)} \left[M^0\right]_{(n-1)\times 1}. \tag{4.29}$$

Now, calculate the formula by using the values of $\mathcal{Q}^0$ and h^0 from eqn (4.27) and (4.29), respectively.

$$\mathcal{Q}^1 = \mathcal{Q}^0 + h^0. \quad (4.30)$$

The obtained value of $\mathcal{Q}^1$ from eqn (4.30) can be represented as follows:

$$\mathcal{Q}^1 = \begin{bmatrix} \theta_2^1(\xi) \\ \theta_3^1(\xi) \\ \theta_4^1(\xi) \\ \vdots \\ \theta_n^1(\xi) \end{bmatrix}. \quad (4.31)$$

Now the obtained values of $\theta_2^1, \theta_3^1, \theta_4^1, \ldots, \theta_n^1$ in eqn (4.30) are assumed to be the initial values for the next iteration. The procedure is repeated again for the initial values $\theta_2^1, \theta_3^1, \theta_4^1, \ldots, \theta_n^1$ from eqn (4.26) such as h^1 can be calculated as follows:

$$h^1 = -\left[\mathcal{J}^1\right]^{-1}_{(n-1)\times(n-1)} \left[M^1\right]_{(n-1)\times 1}, \quad (4.32)$$

where $\mathcal{J}^1$ and M^1 represent the initial values $\theta_2^1, \theta_3^1, \theta_4^1, \ldots, \theta_n^1$.

$$\mathcal{Q}^2 = \mathcal{Q}^1 + h^1. \quad (4.33)$$

Repeat the process until the convergence is achieved as defined below:

$$h^m = -\left[\mathcal{J}^m\right]^{-1}_{(n-1)\times(n-1)} \left[M^m\right]_{(n-1)\times 1} \quad (4.34)$$

$$Q^{m+1} = Q^m + h^m, \quad (4.35)$$

where $m = 0, 1, 2, 3, \ldots$ Continue the above procedure till numerical values $\theta(\xi)$ change negligibly or do not change by calculating further iterations.

4.3 Convergence Analysis

To check the convergence of the numerical results, it is convenient to use norms. To know the accuracy of the approximate solution obtained from the proposed method based on finding the norms such as $\|\theta\|_2$ and $\|\theta\|_\infty$ norms and the error analysis based on absolute error, $\|\theta - \theta_h\|_2$ and $\|\theta - \theta_h\|_\infty$, where θ represents the analytic solution and θ_h represents the approximate solution [10].

L_p norm with the value of p such that $1 \leq p < \infty$ is said to converge to the exact solution if $\|\theta - \theta_h\|_p \longrightarrow 0$ as $n \longrightarrow \infty$.

In this chapter, L_2 and L_∞ norms have been calculated by using weight functions [9, 12]:

$$\|\theta\|_2 = \sqrt{\sum_{i=1}^{n+1} w_i(\xi)\theta_i^2},$$

and the L_2 norm can be calculated as follows:

$$\|\theta - \theta_h\|_2 = \sqrt{\sum_{i=1}^{n+1} w_i(\xi)(\theta - \theta_h)_i^2}.$$

L_∞ norm is also known as the maximum norm and can be written as

$$\|\theta\|_\infty = \max |\theta_i|; \, i = 1, 2, 3, ..., n+1,$$

and to check the maximum error, the norm can be calculated as

$$\|\theta - \theta_h\|_\infty = \max |(\theta - \theta_h)_i|; \quad i = 1, 2, 3, ..., n+1.$$

4.4 Numerical Examples

Example 1: Consider a linear Lane–Emden equation

$$\frac{d^2\theta(\xi)}{d\xi^2} + \frac{1}{\xi}\left(\frac{d\theta(\xi)}{d\xi}\right) = \left(\frac{8}{8-\xi^2}\right)^2$$

including boundary conditions $\theta(1) = 0$ and $\theta'(0) = 0$ with the exact solution $\theta(\xi) = 2\log\left(\frac{7}{8-\xi^2}\right)$ [4]. In Table (4.1), numerical values have been compared to the exact values in terms of absolute error. A good comparison has been observed between the two, and the absolute error is found to be of order 10^{-10}. The comparison of results with the exact solutions and with the solutions obtained by the B-spline approach [4] shows that the results obtained by using the Bessel collocation method are better than the B-spline method.

In Figure (4.1), comparison of numerical and exact values has been shown graphically and a good match has been found between the two.

4.4 Numerical Examples

Table 4.1 Comparison of numerical values from the Bessel collocation method and B-spline method in terms of absolute error.

Collocation points	Exact solution	Numerical solution by BCM	Absolute error by BCM	Absolute error by B-spline method [4]	Absolute error by [18]
0.1	−0.26456122145	−0.26456122137	7.8420e-11	5.7786e-06	−
0.2	−0.25703770160	−0.25703770155	5.6279e-11	2.9840e-07	3.8810e-09
0.3	−0.24443526545	−0.24443526576	3.1375e-10	5.7346e-06	−
0.4	−0.22665737061	−0.22665737059	1.9153e-11	5.6294e-06	3.9877e-09
0.5	−0.20356538862	−0.20356538827	3.4701e-10	4.6114e-06	−
0.6	−0.17497490825	−0.17497490836	1.0925e-10	4.0918e-06	4.0502e-09
0.7	−0.14065063344	−0.14065063379	3.4585e-10	3.3666e-06	−
0.8	−0.10029956737	−0.10029956718	1.8948e-10	2.4326e-06	4.0959e-09
0.9	−0.05356204535	−0.05356204532	3.1992e-11	9.5464e-07	−
L_2	−	−	7.2475e-11	−	−
L_∞	−	−	1.1853e-10	−	−

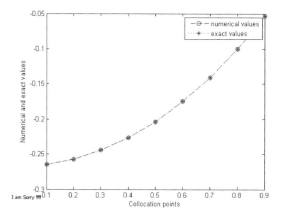

Figure 4.1 Graphical representation to the Example 1.

Example 2: Consider the nonlinear Lane–Emden equation

$$\frac{d^2\theta(\xi)}{d\xi^2} + \frac{1}{\xi}\left(\frac{d\theta(\xi)}{d\xi}\right) + e^{\theta(\xi)} = 0$$

with boundary conditions $\theta'(0) = 0, \theta(1) = 0$. The exact solution to this problem is $\theta(\xi) = 2\ln\left(\frac{\alpha+1}{\alpha\xi^2+1}\right)$ where $\alpha = 3 - 2\sqrt{2}$ [2].

The absolute errors, L_2 norm, and L_∞ norm with respect to the weight function have been calculated. In Table (4.2), the values of θ obtained using the Bessel collocation method have been compared to the exact values at node points. The absolute error is found to be of order 10^{-6}. In Figure (4.2), numerical values have been compared to the exact values graphically and a good match has been observed between exact and numerical values.

Table 4.2 Comparison of exact and numerical values at node points.

Collocation points	Exact solution	Numerical solution by BCM	Absolute error by BCM
0.1	0.313269401	0.313265850	3.5502e-06
0.2	0.303017975	0.303015423	2.5519e-06
0.3	0.286046949	0.286047265	3.1677e-07
0.4	0.262528148	0.262531127	2.9790e-06
0.5	0.232692979	0.232696784	3.8053e-06
0.6	0.196824158	0.196826806	2.6480e-06
0.7	0.155247456	0.155248107	6.5095e-07
0.8	0.108323430	0.108322763	6.6700e-07
0.9	0.056439164	0.056438603	5.6162e-07
L_2	–	–	1.7341e-06
L_∞	–	–	3.3788e-06

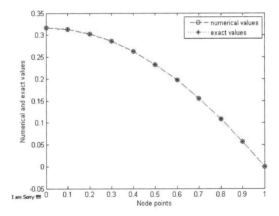

Figure 4.2 Graphical representation to the Example 5.

Example 3: Consider a nonlinear Lane–Emden differential equation

$$\frac{d^2\theta(\xi)}{d\xi^2} + \frac{2}{\xi}\left(\frac{d\theta(\xi)}{d\xi}\right) - 6\theta(\xi) - 4\theta(\xi)\ln(\theta(\xi)) = 0; 0 \leq \xi \leq 1$$

including boundary conditions $\theta(0) = 1$ and $\theta'(0) = 0$. This problem has exact solution $\theta(\xi) = \exp(\xi^2)$ [2]. In Table (4.3), the numerical values have been compared to the exact values by calculating absolute error at different collocation points. In Figure (4.3), numerical values obtained from the Bessel collocation method has been compared to exact values and a good match has been observed between the two.

Table 4.3 Comparison of exact and numerical values at node points.

Collocation points	Exact solution	Numerical solution by BCM	Absolute error by BCM
0.1	1.01005013	1.01005017	3.6909e-08
0.2	1.04080766	1.04081077	3.1108e-06
0.3	1.09413785	1.09417428	3.6427e-05
0.4	1.17329862	1.17351087	2.1225e-04
0.5	1.28317623	1.28402541	8.4918e-04
0.6	1.43064131	1.43332941	2.6881e-03
0.7	1.62505284	1.63231622	7.2634e-03
0.8	1.87894995	1.89648088	1.7531e-02
0.9	2.20898683	2.24790798	3.8921e-02
L_2	–	–	8.1095e-02
L_∞	–	–	3.3856e-02

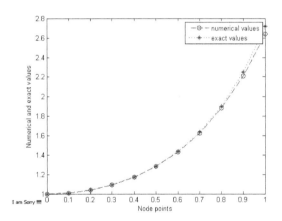

Figure 4.3 Graphical representation to the Example 6.

4.5 Conclusions

The Bessel collocation method with Legendre collocation points has been successfully implemented on three Lane–Emden equations of linear and non-linear type. Rate of convergence has been checked by calculating Euclidean and maximum norms. Numerical values obtained from the Bessel collocation method have been compared with the B-spline method and are found to be better than the latter. By calculating absolute error, $\|\theta - \theta_h\|_\infty$ and $\|\theta - \theta_h\|_2$ with respect to the weight function, it is found that the numerical approach is stable and the results obtained by this approach are consistent and convergent.

References

[1] W. N. Bailey. *Generalized Hypergeometric Series*. Hafner, New York, 1972.

[2] A. H. Bhrawy and A. S. Alofi. A jacobi-gauss collocation method for solving nonlinear lane-emden type equation. *Commun. Nonlinear Sci.*, 17.

[3] J. P. Boyd. Chebyshev spectral methods and the lane-emden problem. *Numer. Math. Theor. Method. Appl.*, 4.

[4] N. Caglar and H. Caglar. B-spline solution of singular boundary value problems. *Appl. Math. Comput.*, 182.

[5] W. N. Everitt and C. Markett. On a generalization of bessel functions satisfying higher-order differential equations. *J. Comput. Appl. Math.*, 54.

[6] N. B. Ferguson and B. A. Finlayson. Finlayson, transient chemical reaction analysis by orthogonal collocation. *Chem. Eng J.*, 1(4):327–336, 1970.

[7] B. A. Finlayson. Packed bed reactor analysis by orthogonal collocation. *Chem. Eng. Sci.*, 26.

[8] W. Gautschi. *Numerical Analysis*. Second Edition, Springer-Verlag, New York, 2012.

[9] T. H. Koornwinder. Orthogonal polynomial with weight function $(1-x)^\alpha(1-x)^\beta + m\delta(x-1) + n\delta(x-1)$. *Can. Math. Bull.*, 27(2):205–214, 1984.

[10] N. W. McLACHLAN. *Bessel functions for engineers*. Second Edition, Oxford University Press, 1961.

[11] M. A. Noor and M. Waseem. Some iterative method for solving a system of nonlinear equations. *Comput. Math. Appl.*, 57.

[12] A. K. Pani P. Mishra, K. K. Sharma and G. Fairweather. Orthogonal spline collocation for singularly perturbed reaction diffusin problems in one dimention. *Int. J. Numer. Anal. Mod.*, 16.

[13] K. Paranda and M. Hemamia. Numerical study of astrophysics equations by meshless collocation method based on compactly supported radial basis function. *Int. J. Comput. Math.*, 3.

[14] S. L. Ross. *Differential Equations*. Third Eddition, Uniersity of New Hampshire, 1984.

[15] S. S. Dhaliwal S. Arora and V. K. Kukreja. Solution of two point boundary value problems using orthogonal collocation on finite elements. *Appl. Math. Comput.*, 171.

[16] I. N. Sneddon. *Special function of mathematical physics and chemistry*. Third Eddition, Longman Mathematical Texts, 1980.
[17] Ş. Yüzbaşi. Bessel collocation approach for solving continuous population models for single and interacting species. *Appl. Math. Model*, 36.
[18] Ş. Yüzbaşi. An improved bessel collocation method with a residual error function to solve a class of lane–emden differential equations. *Math. Comput. Modelling*, 57.
[19] Ş. Yüzbaşi. A numerical approximation based on the bessel functions of first kind for solutions of riccati type differential-difference equations. *Comput. Math. Appl.*, 64(6):1691–1705, 2012.
[20] J. V. Villadsen and W. E. Stewart. Solution of boundary value problem by orthogonal collocation. *Chemical Engineering Science*, 22.
[21] G. N. Watson. *The theory of Bessel functions*. Cambridge University Press, Cambridge, 1994.

5

B-spline Basis Function and its Various Forms Explained Concisely

Shubham Mishra[1], Geeta Arora[1], and Homan Emadifar[2]

[1]Department of Mathematics, Lovely Professional University, India
[2]Department of Mathematics, Hamedan Branch,
Islamic Azad University, Iran
E-mail: mishrasam73@gmail.com; geetadma@gmail.com; homan_emadi@yahoo.com

Abstract

Whenever the numerical solution to a set of differential equations is required, whether ordinary or partial differential equations, it needs to express in terms of some basis functions. One of the basis functions that have all the properties required for the solution of differential equations is B-spline. In this work, the applications of B-spline basis function and its implementation on different methods are presented, which include differential quadrature method, finite element method, collocation method, and many more. Other than the methods, the boundary value problems that are solved by the B-spline basis function in literature by different researchers are also discussed. To the authors' knowledge, there is no review work of B-spline basis function reported in the literature. Hence, this work is beneficial for the researchers to get acquainted with each and every prerequisite required to work with B-spline basis function.

Keywords: spline, B-spline, cubic spline, quintic spline, quadratic spline, quartic spline.

5.1 Introduction

Mathematical models of most of the science and engineering problems and phenomena are expressed in terms of differential equations, either ordinary or partial, and as a system of initial and boundary value problems either linear or nonlinear. These equations have been intensively investigated in the literature as per its applicability in engineering and distinct areas such as quantum mechanics, electromagnetic fields, fluid flow diffusion, etc. Since finding an analytical solution to these equations is difficult, advanced numerical methods must be utilized. There are a variety of numerical techniques used for solving partial differential equations that include collocation method, differential quadrature method, and finite element method.

5.1.1 Idea of spline

Consider a problem of finding a polynomial that passes through the points whose function values are given. For only two points, there exists a linear polynomial, but if the number of points doubled, a cubic polynomial can be fitted and so on. Thus, with increase in the number of data points, the degree of polynomial also increases. Thus, with a large number of points, a higher degree of polynomial emerges, which is difficult to work with. This situation can be tackled with the use of piecewise polynomial. A piecewise polynomial is a polynomial that approximates the function over some part of the domain. This approximation allows us to construct an exact approximation, but since the portion of approximated polynomials is not smooth, the obtained function is not smooth at the point where the two piecewise polynomials are joined. Because every polynomial may or may not be differentiable, but it is always continuous over the interval, the interpolant graph may not always be smooth. To resolve this problem, spline functions are used.

5.2 B-spline

When it comes to numerical aspects, in the subject of approximation theory, for the purpose of addressing boundary value problems and partial differential equations, the B-spline basis function plays a key role. In the realm of mathematics, Schoenberg [1] developed B-spline ("B" stands for basis) in 1946, characterizing a uniform piecewise polynomial approximation. The term "B-spline" is an abbreviation for "basis spline." A least supported spline function in terms of degree, smoothness, and domain partition is B-spline.

The knot w_i sequence is used in order to define the basis function. Let W be a set consisting of $M + 1$ non-decreasing real numbers

$w_0 \leq w_1 \leq w_2 \leq \cdots \leq w_{M-1} \leq w_M$. The set W signifies the working region of the to define the B-spline basis; the ith knot span is defined by the half-open interval $[w_i, w_{i+1})$ and the knot series represents the working region of the real number line.

The knot vectors or the knot series are called **uniform** if the knots are uniformly distributed (i.e., $w_{i+1} - w_i$ is a constant for $0 \leq i \leq M-1$); otherwise, it is said to be **non-uniform**. Having degree r, each B-spline function covers $r+1$ knots or r intervals. Schumaker [2] expressed the B-spline basis functions based on the concept of divided differences. Cox [3] and Boor separately discovered a recurrence relation for calculating B-spline basis functions, in the 1970s. The following is the discussed formula for the pth B-spline basis function having rth degree proposed by Boor in a recursive way by utilizing the Leibniz theorem:

$$B_{p,r}(w) = V_{p,r} B_{p,r-1}(w) + (1 - V_{p+1,r}) B_{p+1,r-1}(w), \text{ for } r \geq 1 \quad (5.1)$$

$$\text{Here, } V_{p,r} = \left(\frac{w - w_r}{w_{p+r} - w_r} \right)$$

The Cox–de Boor recursion formula is expressed as in the above form. Here, $B_{p,r}(w)$ defines a pth B-spline basis function having degree r. By using this formula, it is shown that B-spline basis functions of any degree can be defined as a linear combination of basis functions of lower degrees.

The value of zero-degree B-spline is 1 in a half-open internal, and, otherwise, it is 0, which is defined as follows.

By putting $r=1$ in the recursive formula using zero-degree B-spline formula. **B-spline** having **first degree** is also called **linear B-spline**.

The value of the function obtained for linear B-spline can be represented as follows:

$$B_{i,1} = \begin{cases} \frac{w-a}{b-a} & w \in [a, b) \\ \frac{b-a}{c-b} & w \in [b, c) \\ 0 & \text{otherwise} \end{cases} \quad (5.2)$$

where the points are discretized as $w_i = a$, $w_{i+1} = b$, $w_{i+2} = c$, and so on.

Using the value of **linear B-spline** and for $r = 2$ in the recursive formula, the B-spline basis function of **second degree** is given as follows:

$$B_{i,2} = \begin{cases} \frac{(w-a)^2}{(c-b)(b-a)} & w \in [a, b) \\ \frac{(w-a)(c-w)}{(c-a)(c-b)} + \frac{(c-w)(w-b)}{(d-b)(c-b)} & w \in [b, c) \\ \frac{(d-w)^2}{(d-b)(d-c)} & w \in [c, d) \\ 0 & \text{otherwise} \end{cases} \quad (5.3)$$

Table 5.1 At the nodal points, values of $B_i(w)$ for cubic B-spline and its derivatives.

	q	p	a	b	c
$B_i(w)$	0	1	4	1	0
$B_i'(w)$	0	u	0	$-u$	0
$B_i''(w)$	0	$\frac{2}{3}u^2$	$-\frac{4}{3}u^2$	$\frac{2}{3}u^2$	0

The **cubic B-spline basis function** is third degree B-spline and is given by the following formula:

$$B_{i,3} = \frac{1}{h^3} \begin{cases} (w-q)^3 & w \in [q,p) \\ (w-q)^3 - 4(w-p)^3 & w \in [p,a) \\ (c-w)^3 - 4(b-w)^3 & w \in [a,b) \\ (c-w)^3 & w \in [b,c) \\ 0 & \text{otherwise} \end{cases} \quad (5.4)$$

From the definition given by eqn (4), at the nodal points, the values of $B_i(w)$ can be obtained. On differentiating with respect to w, the first and second derivative values of $B_i(w)$ can be obtained. Table 5.1 provides the value, $u = \frac{3}{h}$, at the nodal points, the values of $B_i(w)$ and its derivatives.

The **fourth degree B-spline basis function** known as **quartic B-spline** is given by

$$B_{i,4} = \frac{1}{h^4} \begin{cases} (w-q)^4 & w \in [q,p) \\ (w-q)^4 - 5(w-p)^4 & w \in [p,a) \\ (w-q)^4 - 5(w-p)^4 + 10(w-a)^4 & w \in [a,b) \\ (d-w)^4 - 5(c-w)^4 & w \in [b,c) \\ (d-w)^4 & w \in [c,d) \\ 0 & \text{otherwise} \end{cases}$$

(5.5)

This basis function is non-zero on five knot spans. From the definition given by eqn (5.5), the values of $\boldsymbol{B_i}(\boldsymbol{w})$ at the nodal points can be obtained. On differentiating with respect to w, its three derivative values can be obtained in an identical approach. Using the value $= \frac{4}{h}$, at the nodal points, the value of $\boldsymbol{B_i}(\boldsymbol{w})$ and its derivatives may be tabulated as in Table 5.2.

Extending the definition, the quintic B-spline basis function can be defined as follows:

5.2 B-spline

Table 5.2 At the nodal points, value of $B_i(w)$ for quartic B-spline and its derivatives.

w	q	p	a	b	c	d
$B_i(w)$	0	1	11	11	1	0
$B_i'(w)$	0	v	$3v$	$-3v$	$-3v$	0
$B_i''(w)$	0	$\frac{3}{4}v^2$	$-\frac{3}{4}v^2$	$-\frac{3}{4}v^2$	$\frac{3}{4}v^2$	0
$B_i'''(w)$	0	$\frac{3}{8}v^3$	$-\frac{9}{8}v^3$	$\frac{9}{8}v^3$	$-\frac{3}{8}v^3$	0

Table 5.3 At the nodal points, value of $B_i(w)$ for quintic B-spline and its derivatives.

w	r	q	p	a	b	c	d
$B_i(w)$	0	1	26	66	26	1	0
$B_i'(w)$	0	ω	10ω	0	-10ω	$-\omega$	0
$B_i''(w)$	0	$\frac{4}{5}\omega^2$	$\frac{8}{5}\omega^2$	$-\frac{24}{5}\omega^2$	$\frac{8}{5}\omega^2$	$\frac{4}{5}\omega^2$	0
$B_i'''(w)$	0	$\frac{12}{25}\omega^3$	$-\frac{24}{25}\omega^3$	0	$\frac{24}{25}\omega^3$	$-\frac{12}{25}\omega^3$	0
$B_i^{iv}(w)$	0	$\frac{24}{125}\omega^4$	$-\frac{96}{125}\omega^4$	$\frac{144}{125}\omega^4$	$-\frac{96}{125}\omega^4$	$\frac{24}{125}\omega^4$	0

$$B_{i,5}(w) = \frac{1}{h^5}\begin{cases} (w-r)^5 & w \in [r,q) \\ (w-r)^5 - 6(w-q)^5 & w \in [q,p) \\ (w-r)^5 - 6(w-q)^5 + 15(w-p)^5 & w \in [p,a) \\ (d-w)^5 - 6(c-w)^5 + 15(b-w)^5 & w \in [a,b) \\ (d-w)^5 - 6(c-w)^5 & w \in [b,c) \\ (d-w)^5 & w \in [c,d) \\ 0 & \text{otherwise} \end{cases} \quad (5.6)$$

The definition provides the values of $B_{i,5}(w)$ at different values of nodes. In an identical approach, the value of its all four derivatives can be obtained on differentiating with respect to w. Table 5.3 provides the value of derivatives considering $v = \frac{5}{h}$.

The degree of B-spline can be extended as sixth degree, seventh degree, eighth degree, ninth degree, and so on. The use of B-spline up to tenth degree exists in literature. Other than the standard B-spline, there are different types of B-splines that can be used according to the considered function such as trigonometric B-spline, exponential B-spline, and so on.

The third-degree B-spline is the most frequently used degree for the B-spline basis function; so, in this work, the author focuses on the third-degree basis function in various forms.

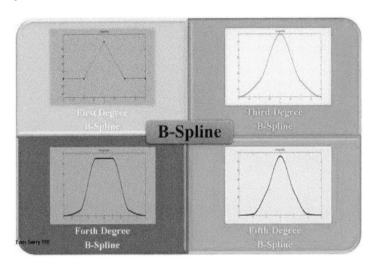

5.2.1 Trigonometric B-spline

A trigonometric B-spline $T_i(w)$ is characterized as a spline function that has minimum supports for a given degree, smoothness, and space partition. Having degree r, each trigonometric B-spline function covers $r + 1$ knots or r intervals and is defined as follows [17, 28]:

$$T_i^r(w) = \frac{\sin\frac{w-w_i}{2}}{\sin\frac{w_{i+r-1}-w_i}{2}} T_i^{r-1}(w) + \frac{\sin\frac{w_{i+r}-w}{2}}{\sin\frac{w_{i+r}-w_{i+1}}{2}} T_{i+1}^{r-1},$$

for $r = 2, 3, 4, \ldots$.

This recurrence relation demonstrates that the trigonometric B-spline basis elements of a higher degree can be steadily assessed using the basis of lower degree. $T_i(w)$ is a piecewise trigonometric function with geometric properties like non-negativity, partition of unity, and C^∞ continuity.

5.2.1.1 Three degree or cubic trigonometric B-spline

The cubic trigonometric B-spline basis function $T_i(w)$, for $i = -1, 0, \ldots, M+1$, can be expressed in the following form [19]:

$$T_i^3(w) = \frac{1}{\omega} \begin{cases} p^3(w_i), & w \in [w_i, w_{i+1}) \\ p(w_i)(p(w_i)q(w_{i+2}) + q(w_{i+3})p(w_{i+1})) \\ \quad + q(w_{i+4})p^2(w_{i+1}), & w \in [w_{i+1}, w_{i+2}) \\ q(w_{i+4})(p(w_{i+1})q(w_{i+3}) + q(w_{i+4})p(w_{i+2})) \\ \quad + p(w_i)q^2(w_{i+3}), & w \in [w_{i+2}, w_{i+3}) \\ q^3(w_{i+4}), & w \in [w_{i+3}, w_{i+4}) \\ 0, & \text{otherwise} \end{cases},$$

(5.7)

where $p(w_i) = \sin\left(\frac{w-w_i}{2}\right), q(w_i) = \sin\left(\frac{w-w_i}{2}\right), \omega = \sin\left(\frac{h}{2}\right)\sin(h)\sin\left(\frac{3h}{2}\right)$, and $h = \frac{b-a}{n}$.

5.2.2 Hyperbolic B-spline

The hyperbolic B-spline HB_i^r [5], of order r, in relation to the partition W can be defined as follows:

$$\text{HB}_i^1(w) = \begin{cases} 1, & \text{when } w_i \leq w < w_{i+1} \\ 0, & \text{otherwise} \end{cases}$$

and for $r > 1$,

$$\text{HB}_i^r(w) = \frac{s(w-w_i)}{s(w_{r+i-1}-w_i)}\text{HB}_i^{r-1}(w) + \frac{s(w_{i+r}-w)}{s(w_{i+r}-w_{i+1})}\text{HB}_{i+1}^{r-1}(w) \quad (5.8)$$

where $s(w) = \sinh(w)$, {sine hyperbolic function of w}.

The equations given by eqn (5.7) and (5.8) occupy the following properties:

(i) For $r \geq 2$, $\text{HB}_i^r \in C^{r-2}(w)$.
(ii) $\text{HB}_i^r(w)$ is a piecewise hyperbolic function.
(iii) $\text{HB}_i^r(w) \geq 0$.
(iv) Support of $\text{HB}_i^r(w) = [w_i, w_{i+r}]$.
(v) $\text{HB}_i^r \in_r$.

Here, r is the space of the hyperbolic polynomial of order r.

$$\Gamma_r = \begin{cases} \text{span}\left\{\{\sinh(2lw), \cosh(2lw)\}_{l=1}^{\left[\frac{r-1}{2}\right]} \cup \{1\}\right\}, & \text{where } s \text{ is odd} \\ \text{span}\left\{\{\sinh((2l-1)w), \cosh((2l-1)w)\}_{l=1}^{\left[\frac{r}{2}\right]}\right\}, & \text{where } s \text{ is even} \end{cases}.$$

5.2.2.1 Cubic hyperbolic B-spline

The hyperbolic B-spline HB_i^r is defined for node points w_i on a uniform distribution for $a = w_1 < w_2 < w_3 < \cdots < w_n = b$ having order r. Therefore, the cubic hyperbolic B-spline forms the basis for all domain functions $[a, b]$. Following is how it can be presented mathematically [5]:

$$\mathrm{HB}_i^4(w) = \begin{cases} \left[\dfrac{[\sinh(w-w_i)]^3}{\sinh(w_{i+3}-w_i)\ \sinh(w_{i+2}-w_i)\ \sinh(w_{i+1}-w_i)}\right], \\ \qquad w \in [w_i, w_{i+1}] \\ \left[\begin{array}{l}\dfrac{[\sinh(w-w_i)]^2[\sinh(w_{i+2}-w)]}{\sinh(w_{i+3}-w_i)\ \sinh(w_{i+2}-w_i)\ \sinh(w_{i+1}-w_i)} + \\ \dfrac{\sinh(w-w_i)\ \sinh(w_{i+3}-w)\ \sinh(w-w_{i+1})}{\sinh(w_{i+3}-w_i)\ \sinh(w_{i+3}-w_{i+1})\ \sinh(w_{i+2}-w_{i+1})} + \\ \dfrac{[\sinh(w_{i+4}-w)]^2 \sinh(w-w_{i+1})}{\sinh(w_{i+4}-w_{i+1})\ \sinh(w_{i+4}-w_{i+2})\ \sinh(w_{i+3}-w_{i+2})}\end{array}\right], \\ \qquad w \in [w_{i+1}, w_{i+2}] \\ \left[\begin{array}{l}\dfrac{[\sinh(w-w_i)][\sinh(w_{i+3}-w)]^2}{\sinh(w_{i+3}-w_i)\ \sinh(w_{i+3}-w_{i+1})\ \sinh(w_{i+3}-w_{i+2})} + \\ \dfrac{\sinh(w_{i+4}-w)\ [\sinh(w-w_{i+1})]\ [\sinh(w_{i+3}-w)]}{\sinh(w_{i+4}-w_{i+1})\ \sinh(w_{i+3}-w_{i+1})\ \sinh(w_{i+3}-w_{i+2})} + \\ \dfrac{[\sinh(w_{i+4}-w)]^2[\sinh(w-w_{i+2})]}{\sinh(w_{i+4}-w_{i+1})\ \sinh(w_{i+4}-w_{i+2})\ \sinh(w_{i+3}-w_{i+2})}\end{array}\right], \\ \qquad w \in [w_{i+2}, w_{i+3}] \\ \left[\dfrac{[\sinh(w_{i+4}-w)]^3}{\sinh(w_{i+4}-w_{i+1})\ \sinh(w_{i+4}-w_{i+2})\ \sinh(w_{i+3}-w_{i+2})}\right], \\ \qquad w \in [w_{i+3}, w_{i+4}] \\ 0 \qquad \text{otherwise.} \end{cases}$$

(5.9)

5.2.3 Uniform algebraic trigonometric tension B-spline

For the region $[a, b]$ with uniform length h by knots $w_i = a+ih, i = 0, 1, 2, 3, \ldots, M$, the UAT tension B-spline having order 2 can be presented as in the following expression [38]:

$$B_{i,2}(w) = \begin{cases} \dfrac{\sin[\tau(w-w_{i-2})]}{\sin(\tau h)}, & [w_{i-2}, w_{i-1}) \\ \dfrac{\sin[\tau(w_i-w)]}{\sin(\tau h)}, & [w_{i-1}, w_i) \\ 0, & \text{else} \end{cases},$$

where $\tau = \sqrt{\eta}$ (η is a positive real number) for $r \geq 3$, and $B_{i,r}$ is recursively defined as follows:

$$B_{i,r}(w) = \int_{-\infty}^{w} (\delta_{i,r-1} B_{i,r-1}(w) - \delta_{i+1,r-1} B_{i+1,r-1}(w))\, dw$$

and $\delta_{i,j} = \left(\int_{-\infty}^{\infty} B_{i,j}(w)\, dw\right)^{-1}$.

5.2 B-spline

UAT tension B-spline of third order can be expressed as follows:

$$B_{i,3}(w) = \begin{cases} \left[\frac{2\delta_{i2}}{\tau\sin(\tau h)}\right]\left[\sin^2\left\{\frac{\tau(w-w_{i-2})}{2}\right\}\right], & [w_{i-2}, w_{i-1}) \\ 1 - \left[\frac{2\delta_{i2}}{\tau\sin(\tau h)}\right]\left[\sin^2\left\{\frac{\tau(w-w_i)}{2}\right\}\right] \\ \quad - \left[\frac{2\delta_{i+12}}{\tau\sin(\tau h)}\right]\left[\sin^2\left\{\frac{\tau(w-w_{i-1})}{2}\right\}\right], & [w_{i-1}, w_i) \\ \left[\frac{2\delta_{i+12}}{\tau\sin(\tau h)}\right]\left[\sin^2\left\{\frac{\tau(w-w_{i+1})}{2}\right\}\right], & [w_i, w_{i+1}) \\ 0, & \text{elsewhere} \end{cases}$$

Similarly, UAT tension B-spline can be extended for order 4:

$$B_{i,4}(w) =$$

$$\begin{cases} \frac{\delta_{i,3}\delta_{i,2}}{\tau\sin(\tau h)}\left[(w-w_{i-2}) - \frac{\sin[\tau(w-w_{i-2})]}{\tau}\right], & [w_{i-2}, w_{i-1}) \\ \delta_{i,3}\left[\frac{\delta_{i,2}}{\tau\sin(\tau h)}\left\{(w_{i-1}-w_{i-2}) - \frac{\sin[\tau(w_{i-1}-w_{i-2})]}{\tau}\right\}\right. \\ \quad + (w-w_{i-1}) \\ \quad - \frac{\delta_{i,2}}{\tau\sin(\tau h)}\left\{(w-w_{i-1}) - \frac{1}{\tau}\left(\sin(\tau(w-w_i)) + \sin(\tau(w_i-w))\right)\right\} \\ \quad \left. - \frac{\delta_{i+1,2}}{\tau\sin(\tau h)}\left\{(w-w_{i-1}) - \frac{\sin(\tau(w-w_{i-1}))}{\tau}\right\}\right] \\ - \\ \frac{\delta_{i+1,3}\delta_{i+1,2}}{\tau\sin(\tau h)}\left\{(w-w_{i-1}) - \frac{\sin(\tau(w-w_{i-1}))}{\tau}\right\}, & [w_{i-1}, w_i) \\ 1 - \frac{\delta_{i,3}\delta_{i+1,2}}{\tau\sin(\tau h)}\left\{(w_{i+1}-w) + \frac{\sin(\tau(w-w_{i+1}))}{\tau}\right\} \\ - \delta_{i+1,3}\left[\frac{\delta_{i+1,2}}{\tau\sin(\tau h)}\left\{(w_i-w_{i-1}) - \frac{\sin(\tau(w_i-w_{i-1}))}{\tau}\right\}\right. \\ \quad + (w-w_i) - \frac{\delta_{i+1,2}}{\tau\sin(\tau h)} \\ \quad \left\{(w-w_i) - \frac{(\sin(\tau(w-w_{i+1})) - \sin(\tau(w_i-w_{i+1})))}{\tau}\right\} \\ \quad \left. - \frac{\delta_{i+2,2}}{\tau\sin(\tau h)}\left\{(w-w_i) - \frac{\sin(\tau(w-w_i))}{\tau}\right\}\right], & [w_i, w_{i+1}) \\ \frac{\delta_{i+1,3}\delta_{i+2,2}}{\tau\sin(\tau h)}\left[(w_{i+2}-w) + \frac{\sin(\tau(w-w_{i+2}))}{\tau}\right], & [w_{i+1}, w_{i+2}) \\ 0, \text{ otherwise} \end{cases}$$

(5.10)

At the node points, detailed values of $B_{i,4}(w)$ and $B'_{i,4}(w)$:

$$p = \left[\frac{1}{4h\left[\sin^2\left(\frac{\tau h}{2}\right)\right]}\right]\left[h - \frac{\sin(\tau h)}{\tau}\right]$$

$$q = 1 - \left[\frac{1}{2h\left[\sin^2\left(\frac{\tau h}{2}\right)\right]}\right]\left[h - \frac{\sin(\tau h)}{\tau}\right]$$

$$r = \left[\frac{1}{4h\left[\sin^2\left(\frac{\tau h}{2}\right)\right]}\right]\left[h - \frac{\sin(\tau h)}{\tau}\right]$$

$$s = \frac{1}{2h},\ t = \frac{-1}{2h}.$$

Evaluation of $\delta_{i,2}$ and $\delta_{i,3}$:

$$\delta_{i,2} = \frac{\tau \sin(\tau h)}{2\left[\sin^2\left\{\frac{\tau(w_{i-1} - w_{i-2})}{2}\right\} + \sin^2\left(\frac{\tau(w_i - w_{i-1})}{2}\right)\right]},$$

where $\{B_0(w), B_1(w), \ldots, B_M(w), B_{M+1}(w)\}$ forms a basis over given domain.

5.2.4 Exponential B-spline

For a uniform mesh Γ with the knots w_i defined on $[a, b]$ if $B_i(w)$ is the B-splines at the points of Γ together with knots w_i, $i = -3, -2, -1, M+1, M+2, M+3$ outside the interval $[a, b]$ and a finite support on each of the four consecutive intervals $[w_i + rh, w_i + (r+1)h]_{r=-3}^0$, $i = 0, \ldots, N+2$. The function formula can be expressed as follows [6, 7]:

$$B_i(w) = \begin{cases} b_2\left[(w_{i-2} - w) - \frac{1}{p}(\sinh(p(w_{i-2} - w)))\right] & \text{if } w \in [w_{i-2}, w_{i-1}]; \\ a_1 + b_1(w_i - w) + c_1 e^{p(w_i - w)} + d_1 e^{-p(w_i - w)} & \text{if } w \in [w_{i-1}, w_i]; \\ a_1 + b_1(w - w_i) + c_1 e^{p(w - w_i)} + d_1 e^{-p(w - w_i)} & \text{if } w \in [w_i, w_{i+1}]; \\ b_2\left[(w - w_{i+2}) - \frac{1}{p}(\sinh(p(w - w_{i+2})))\right] & \text{if } w \in [w_{i+1}, w_{i+2}]; \\ 0 & \text{otherwise.} \end{cases} \quad (5.11)$$

Here,

$$p = \max_{0 \leq i \leq N} p_i,\ s = \sinh(ph),\ c = \cosh(ph)$$

$$b_2 = \frac{p}{2(phc - s)},\ a_1 = \frac{phc}{phc - s},\ b_1 = \frac{p}{2}\left[\frac{c(c-1) + s^2}{(phc - s)(1 - c)}\right],$$

$$c_1 = \frac{1}{4}\left[\frac{e^{-ph}(1 - c) + s(e^{-ph} - 1)}{(phc - s)(1 - c)}\right],\ d_1 = \frac{1}{4}\left[\frac{e^{ph}(c - 1) + s(e^{ph} - 1)}{(phc - s)(1 - c)}\right].$$

Table 5.4 Exponential B-spline values.

w	w_{i-2}	w_{i-1}	w_i	w_{i+1}	w_{i+2}
$B_i(w)$	0	$\frac{s-ph}{2(ph-s)}$	1	$\frac{s-ph}{2(phc-s)}$	0
$B_i'(w)$	0	$\frac{p(1-c)}{2(ph-s)}$	0	$\frac{p(c-1)}{2(phc-s)}$	0
$B_i''(w)$	0	$\frac{p^2 s}{2(phc-s)}$	$\frac{-p^2 s}{phc-s}$	$\frac{p^2 s}{2(phc-s)}$	0

$B_i(w)$ is twice continuously differentiable. From Table 5.4, the values of $B_i(w)$, $B_i'(w)$, and $B_i''(w)$ at the knots w are obtained.

5.2.4.1 Exponential cubic B-spline

For the uniform mesh $a = w_0 < w_1 < \cdots < w_M = b$, the exponential cubic B-spline, $B_i(x)$, is presented as a piecewise polynomial function as given below:

$$B_i(w) = \begin{cases} b_2\left((w_{i-2}-w) - \frac{1}{p}(\sinh(p(w_{i-2}-w)))\right) & [w_{i-2}, w_{i-1}], \\ a_1 + b_1(w_i - w) + c_1\exp(p(w_i - w)) \\ + d_1\exp(p(w_i - w)) & [w_{i-1}, w_i], \\ a_1 + b_1(w - w_i) + c_1\exp(p(w - w_i)) \\ + d_1\exp(-p(w - w_i)) & [w_i, w_{i+1}], \\ b_2\left((w-w_{i+2}) - \frac{1}{p}(\sinh(p(w-w_{i+2})))\right) & [w_{i+1}, w_{i+2}], \\ 0 & \text{otherwise.} \end{cases} \quad (5.12)$$

Here,

$$a_1 = \frac{phc}{phc-s}, \quad b_1 = \frac{p}{2}\left(\frac{c(c-1)+s^2}{(phc-s)(1-c)}\right), \quad b_2 = \frac{p}{2(phc-s)},$$

$$c_1 = \frac{1}{4}\left(\frac{\exp(-ph)(1-c)+s(\exp(-ph)-1)}{(phc-s)(1-c)}\right),$$

$$d_1 = \frac{1}{4}\left(\frac{\exp(ph)(c-1)+s(\exp(ph)-1)}{(phc-s)(1-c)}\right),$$

and $s = \sinh(ph)$, $c = \cosh(ph)$, and p is a free parameter.
$\{B_{-1}(w), B_0(w), \ldots, B_{M+1}(w)\}$ is a basis defined over the interval $[a, b]$. Each basis function $B_i(w)$ is continuously differentiable twice. The values of $B_i(w)$, $B_i'(w)$, and $B_i''(w)$ at the knots w_i can be computed from eqn (5.12) and are documented in Table 5.5.

Table 5.5 Values of $B_i(x)$ and its first and second derivatives at the knot points.

w	w_{i-2}	w_{i-1}	w_i	w_{i+1}	w_{i+2}
$B_i(w)$	0	$\frac{s-ph}{2(phc-s)}$	1	$\frac{s-ph}{2(phc-s)}$	0
$B_i'(w)$	0	$\frac{p(1-c)}{2(phc-s)}$	0	$\frac{p(c-1)}{2(phc-s)}$	0
$B_i''(w)$	0	$\frac{p^2s}{2(phc-s)}$	$\frac{-p^2s}{phc-s}$	$\frac{p^2s}{2(phc-s)}$	0

that should be determined in computations.

Table 5.6 $T_m(w)$ and its derivatives at connecting points.

w_m	w_{m-1}	w_m	w_{m+1}	w_{m+2}	w_{m+3}
$T_{m,5}$	0	$r\left(\frac{h^2\tau^2+2p_1-2}{2\tau^2}\right)$	$r\left(\frac{h^2\tau^2-2p_1(h^2\tau^2+1)+2}{2\tau^2}\right)$	$r\left(\frac{h^2\tau^2-2p_1(h^2\tau^2+1)+2}{2\tau^2}\right)$	$r\left(\frac{h^2\tau^2+2p_1-2}{2\tau^2}\right)$	0
$T_{m,5}'$	0	$r\left(\frac{h\tau-p_2}{\tau}\right)$	$r\left(\frac{h\tau-3p_2+2p_1h\tau}{\tau}\right)$	$-r\left(\frac{h\tau-3p_2+2p_1h\tau}{\tau}\right)$	$-r\left(\frac{h\tau-p_2}{\tau}\right)$	0
$T_{m,5}''$	0	$r(1-p_1)$	$r(p_1-1)$	$r(-1+p_1)$	$r(p_1-1)$	0
$T_{m,5}'''$	0	$r(\tau p_2)$	$-r(3\tau p_2)$	$r(3\tau p_2)$	$-r(\tau p_2)$	0

5.2.5 Quartic hyperbolic trigonometric B-spline

For equally separated points $\pi:a=w_0<w_1<\cdots<w_M=b$ and points $[w_m, w_{m+1}]$ with mesh space $h=w_{m+1}-w_m, m=0, 1, \ldots, M-1$. The quartic hyperbolic–trigonometric tension (HTT) B-spline can be expressed in the following form [22]:

$$T_{m,5}(w) = r\begin{cases} \frac{\tau^2(-\hat{z}_{m-2})^2+2z_{m-2}-2}{2\tau^2}, & w_{m-2}\leq w\leq w_{m-1}, \\ \frac{-\tau^2\left(3h^2+6h\hat{z}_{m-2}+2(-\hat{z}_{m-2})^2\right)+2p_1\left(\tau^2(\hat{z}_{m-1})^2-2\right)}{2\tau^2} \\ \quad -\frac{(6z_{m-1}+2z_m-4)}{2\tau^2}, & w_{m-1}\leq w\leq w_m, \\ \frac{\tau^2\left(13h^2+10h\hat{z}_{m-2}+2(-\hat{z}_{m-2})^2\right)+p_1\left(2\tau^2\left(11h^2+10h\hat{z}(m-2)\right)\right)}{2\tau^2(23h^2+14h(\hat{z}_{m-2})+2(-\hat{z}_{m-2})^2)+2p_1\left(\tau^2(\tau^2\hat{z}_{m+2})^2-2\right)} \\ \quad -(2z_{m+1}+6z_{m+2}-4), & w_m\leq w\leq w_{m+1}, \\ \frac{\tau^2(\tau_{m+3})^2+2z_{m+3}-2}{2\tau^2}, & w_{m+1}\leq w\leq w_{m+2}, \\ 0, & w_{m+2}\leq w\leq w_{m+3}, \end{cases}$$

(5.13)

where τ is the tension parameter, $r=2h^2(1-p_1)^{-1}$, $z_{m+j}=\cos(\tau(w_{m+j}-w))$, $\hat{z}_{m+j}=(w_{m+j}-w)$, $p_1=\cos(\tau h)$, and $p_2=\sin(\tau h)$; The value of the functions and its derivatives are tabulated in Table 5.6.

5.2.6 Quintic hyperbolic B-spline

Let an equidistance finite knot w_m with distance h such that $a=w_0<w_1<w_2<\cdots<w_{n-1}<w_n=b$ be the partition of the domain $[a, b]$. With knots at

the points $w_n, n = 0, 1, 2, \ldots, M$, let $\mathrm{TB}_m(w)$ be trigonometric quintic B-splines. The splines are arranged in $\{\mathrm{TB}_{-1}, \mathrm{TB}_0, \mathrm{TB}_1, \ldots, \mathrm{TB}_M, \mathrm{TB}_{M+1}\}$ and on $[a, b]$, which forms the basis for any function.

$\mathrm{TB}_m(w)$, the quintic trigonometric B-spline basis function, characterizes a piecewise quintic trigonometric function with properties such as C^∞ continuity, non-negativity, and partition of unity for $m = -2, -1, 0, \ldots, M+2$. The basis functions can be expressed by the following formula [18]:

$$\mathrm{TB}_m(w) = \frac{1}{w} \begin{cases} p^5(w_m), & \text{if } w \in [w_m, w_{m+1}) \\ p^4(w_m)q(w_{m+2}) + p^3(w_m)q(w_{m+3})p(w_{m+1}) \\ +p^2(w_m)q(w_{m+4})p^2(w_{m+1}) \\ +p(w_m)q(w_{m+5})p^3(w_{m+1}) \\ +q(w_{m+6})p^4(w_{m+1}), & \text{if } w \in [w_{m+1}, w_{m+2}) \\ p^3(w_m)q^2(w_{m+3}) + p^2(w_m)q(w_{m+4})p(w_{m+1})q(w_{m+3}) \\ +p^2(w_m)q^2(w_{m+4})p(w_{m+2}) \\ +p(w_m)q(w_{m+5})p^2(w_{m+1})q(w_{m+3}) \\ +p(w_m)q(w_{m+5})p(w_{m+1})q(w_{m+4})p(w_{m+2}) \\ +p(w_m)q^2(w_{m+5})p^2(w_{m+2}) \\ +q(w_{m+6})p^3(w_{m+1})q(w_{m+3}) + q(w_{m+6}) \\ \times p^2(w_{m+1})q(w_{m+4})p(w_{m+2}) \\ +q(w_{m+6})p(w_{m+1})q(w_{m+5})p^2(w_{m+2}) \\ +q^2(w_{m+6})p^3(w_{m+2}), & \text{if } w \in [w_{m+2}, w_{m+3}) \\ p^2(w_m)q^3(w_{m+4}) + p(w_m)q(w_{m+5})p(w_{m+1})q^2(w_{m+4}) \\ +p(w_m)q^2(w_{m+5})p(w_{m+2})q(w_{m+1}) \\ +p(w_m)q^3(w_{m+5})p(w_{m+3}) + q(w_{m+6})p^2(w_{m+1})q^2(w_{m+4}) \\ +q(w_{m+6})p(w_{m+1}) \\ \times q(w_{m+5})p(w_{m+2})q(w_{m+4}) + q(w_{m+6})p(w_{m+1})q^2(w_{m+5})p(w_{m+3}) \\ +q^2(w_{m+6})p^2(w_{m+2})q(w_{m+4}) \\ +q^2(w_{m+6})p(w_{m+2}) \\ q(w_{m+5})p(w_{m+3}) \\ \times q^3(w_{m+6})p^2(w_{m+3}), & \text{if } w \in [w_{m+3}, w_{m+4}) \\ p(w_m)q^4(w_{m+5}) + q(w_{m+6})p(w_{m+1})q^3(w_{m+5}) \\ +q^2(w_{m+6})p(w_{m+2})q^2(w_{m+5}) \\ +q^3(w_{m+6})p(w_{m+3})q(w_{m+5}) \\ +q^4(w_{m+6})p(w_{m+4}), & \text{if } w \in [w_{m+4}, w_{m+5}) \\ q^5(w_{m+6}), & \text{if } w \in [w_{m+5}, w_{m+6}) \end{cases}$$

(5.14)

Table 5.7 Table for the values of UAH tension B-spline of order 4, i.e., $UAHB_{(i,4)}(w)$ and $UAHB'_{(i,4)}(w)$ at different node points [23].

	$w_{(i-2)}$	$w_{(i-1)}$	w_i	$w_{(i+1)}$	$w_{(i+2)}$
$UAHB(i,4)(w)$	0	b_1	b_2	b_3	0
$UAHB'(i,4)(w)$	0	b_4	0	b_5	0

5.2.7 Modified cubic UAH (uniform algebraic hyperbolic) tension B-spline

By using the following set of equations, improvised values can be obtained.

Uniform algebraic hyperbolic tension B-spline of order 4 is defined as follows:

$$UAHB_{i,4}(w) = \begin{cases} \frac{\delta_{i,3}\delta_{i,2}}{\tau\sinh(\tau h)}\left[(w_{i-2}-w) + \frac{\sinh[\tau(w-w_{i-2})]}{\tau}\right], & [w_{i-2}, w_{i-1}) \\ \delta_{i,3}\left[\frac{\delta_{i,2}}{\tau\sinh(\tau h)}\left\{(w_{i-2}-w_{i-1}) + \frac{\sinh[\tau(w_{i-1}-w_{i-2})]}{\tau}\right\} + (w-w_{i-1})\right. \\ \left. - \frac{\delta_{i,2}}{\tau\sinh(\tau h)}\left\{(w_{i-1}-w) + \frac{1}{\tau}\left(\sinh(\tau(w-w_i)) + \sinh(\tau(w_i-w_{i-1}))\right)\right\}\right. \\ \left. - \frac{\delta_{i+1,2}}{\tau\sinh(\tau h)}\left\{(w_{i-1}-w) + \frac{\sinh(\tau(w-w_{i-1}))}{\tau}\right\}\right] \\ \frac{\delta_{i+1,3}\delta_{i+1,2}}{\tau\sinh(\tau h)}\left\{(w_{i-1}-w) + \frac{\sin(\tau(w-w_{i-1}))}{\tau}\right\}, & [w_{i-1}, w_i) \\ 1 - \frac{\delta_{i,3}\delta_{i+1,2}}{\tau\sinh(\tau h)}\left\{(w-w_{i+1}) - \frac{\sinh(\tau(w-w_{i+1}))}{\tau}\right\} \\ -\delta_{i+1,3}\left[\frac{\delta_{i+1,2}}{\tau\sinh(\tau h)}\left\{(w_{i-1}-w_i) + \frac{\sinh(\tau(w_i-w_{i-1}))}{\tau}\right\}\right. \\ \left. + (w-w_i) - \frac{\delta_{i+1,2}}{\tau\sinh(\tau h)}\right. \\ \left. \left\{(w_i-w) + \frac{(\sinh(\tau(w-w_{i+1})) + \sinh(\tau(w_i-w_{i+1})))}{\tau}\right\}\right. \\ \left. - \frac{\delta_{i+2,2}}{\tau\sinh(\tau h)}\left\{(w_i-w) + \frac{\sinh(\tau(w-w_i))}{\tau}\right\}\right], & [w_i, w_{i+1}) \\ \frac{\delta_{i+1,3}\delta_{i+2,2}}{\tau\sinh(\tau h)}\left[(w-w_{i+2}) - \frac{\sinh(\tau(w-w_{i+2}))}{\tau}\right], & [w_{i+1}, w_{i+2}) \\ 0, & \text{otherwise} \end{cases}$$

(5.15)

the values of the function and its derivative are tabulated in Table 5.7. Similarly, for second- or higher-order derivative, the weighting coefficients b_{ij}^r can be obtained by the following formula:

$$b_{ij}^r = r\left[b_{ij}^1 b_{ii}^{r-1} - \frac{b_{ij}^{r-1}}{y_i - y_j}\right] \text{ for } i \neq j.$$

5.2 B-spline

Table 5.8 Value of UAT tension B-spline at different node points.

	w_{i-2}	w_{i-1}	w_i	w_{i+1}	w_{i+2}
$\mathbf{UAHB}_{i,4}(w)$	0	a_1	a_2	a_3	0
$\mathbf{UAHB}'_{i,4}(w)$	0	a_4	0	a_5	0

Here, $i = 1, 2, 3, \ldots, N$ and $r = 2, 3, 4, \ldots, N - 1$.

$$b_{ii}^r = -\sum_{j=1, j \neq i}^{N} b_{ij}^{(r)} \text{ for } i = j.$$

5.2.8 Modified cubic UAT tension B-spline

UAT tension B-spline with order 4 is defined as follows [24]:

$$\text{UATB}_{i,4}(w) = \begin{cases} \frac{\delta_{i,3}\delta_{i,2}}{\tau\sin(\tau h)}\left[(w-w_{i-2}) - \frac{\sin[\tau(w-w_{i-2})]}{\tau}\right], & [w_{i-2}, w_{i-1}] \\ \delta_{i,3}\left[\frac{\delta_{i,2}}{\tau\sin(\tau h)}\left\{(w_{i-1}-w_{i-2}) - \frac{\sin[\tau(w_{i-1}-w_{i-2})]}{\tau}\right\} + (w-w_{i-1}) \right. \\ \quad -\frac{\delta_{i,2}}{\tau\sin(\tau h)}\left\{(w-w_{i-1}) - \frac{1}{\tau}\left(\sin\left(\tau\left(w-w_i\right)\right) + \sin\left(\tau\left(w_i-w\right)\right)\right)\right\} \\ \quad -\frac{\delta_{i+1,2}}{\tau\sin(\tau h)}\left\{(w-w_{i-1}) - \frac{\sin(\tau(w-w_{i-1}))}{\tau}\right\}\right] \\ \frac{\delta_{i+1,1}\delta_{i+1,2}}{\tau\sin(\tau h)}\left\{(w-w_{i-1}) - \frac{\sin(\tau(w-w_{i-1}))}{\tau}\right\}, & [w_{i-1}, w_i] \\ 1-\frac{\delta_{i,5}\delta_{i+1,2}}{\tau\sin(\tau h)}\left\{(w_{i+1}-w) + \frac{\sin(\tau(w-w_{i+1}))}{\tau}\right\} \\ -\delta_{i+1,3}\left[\frac{\delta_{i+1,2}}{\tau\sin(\tau h)}\left\{(w_i-w_{i-1}) - \frac{\sin(\tau(w_i-w_{i-1}))}{\tau}\right\} \right. \\ \quad + (w-w_i) - \frac{\delta_{i+1,2}}{\tau\sin(\tau h)}\left\{(w-w_i) - \frac{(\sin(\tau(w-w_{i+1}))-\sin(\tau(w_i-w_{i+1})))}{\tau}\right\} \\ \quad -\frac{\delta_{i+2,2}}{\tau\sin(\tau h)}\left\{(w-w_i) - \frac{\sin(\tau(w-w_i))}{\tau}\right\}\right], & [w_i, w_{i+1}] \\ \frac{\delta_{i+1,3}\delta_{i+2,2}}{\tau\sin(\tau h)}\left[(w_{i+2}-w) + \frac{\sin(\tau(w-w_{i+2}))}{\tau}\right], & [w_{i+1}, w_{i+2}] \\ 0, & \text{otherwise} \end{cases}$$

(5.16)

Similarly, for second- or higher-order derivatives, the weighting coefficients a_{ij}^r can be obtained by the following formula:

$$a_{ij}^r = r\left[a_{ij}^1 a_{ii}^{r-1} - \frac{a_{ij}^{r-1}}{y_i - y_j}\right] \text{ for } i \neq j.$$

Here, $i = 1, 2, 3, \ldots, M$ and $r = 2, 3, 4, \ldots, M-1$.

$$a_{ii}^r = -\sum_{j=1, j \neq i}^{N} a_{ij}^{(r)} \text{ for } i = j.$$

5.2.9 Quintic trigonometric B-spline

$\text{TB}_m(w)$, the quintic trigonometric B-spline basis function, for $m = -2, -1, 0, \ldots, M+2$ with properties such as non-negativity, C^∞ continuity, and partition of unity, defines a piecewise quintic trigonometric function. The basis functions can be expressed by the following formula [36]:

$$\text{TB}_m(w) = \frac{1}{\omega} \begin{cases} p^5(w_m), & \text{if } w \in [w_m, w_{m+1}) \\ p^4(w_m) q(w_{m+2}) + p^3(w_m) q(w_{m+3}) p(w_{m+1}) \\ + p^2(w_m) q(w_{m+4}) p^2(w_{m+1}) \\ + p(w_m) q(w_{m+5}) p^3(w_{m+1}) + q(w_{m+6}) p^4(w_{m+1}), & \text{if } w \in [w_{m+1}, w_{m+2}) \\ p^3(w_m) q^2(w_{m+3}) \\ + p^2(w_m) q(w_{m+4}) p(w_{m+1}) q(w_{m+3}) \\ + p^2(w_m) q^2(w_{m+4}) p(w_{m+2}) \\ + p(w_m) q(w_{m+5}) p^2(w_{m+1}) q(w_{m+3}) \\ + p(w_m) q(w_{m+5}) p(w_{m+1}) q(w_{m+4}) p(w_{m+2}) \\ + p(w_m) q^2(w_{m+5}) p^2(w_{m+2}) + q(w_{m+6}) p^3 \\ (w_{m+1}) q(w_{m+3}) + q(w_{m+6}) \\ \times p^2(w_{m+1}) q(w_{m+4}) p(w_{m+2}) + q(w_{m+6}) p(w_{m+1}) \\ q(w_{m+5}) p^2(w_{m+2}) \\ + q^2(w_{m+6}) p^3(w_{m+2}), & \text{if } w \in [w_{m+2}, w_{m+3}) \\ p^2(w_m) q^3(w_{m+4}) \\ + p(w_m) q(w_{m+5}) p(w_{m+1}) q^2(w_{m+4}) \\ + p(w_m) q^2(w_{m+5}) p(w_{m+2}) q(w_{m+1}) \\ + p(w_m) q^3(w_{m+5}) p(w_{m+3}) + q(w_{m+6}) p^2(w_{m+1}) \\ q^2(w_{m+4}) + q(w_{m+6}) p(w_{m+1}) \\ \times q(w_{m+5}) p(w_{m+2}) q(w_{m+4}) + q^2(w_{m+6}) p(w_{m+1}) \\ q(w_{m+5}) p(w_{m+3}) \\ + q^2(w_{m+6}) p^2(w_{m+2}) q(w_{m+4}) + q^2(w_{m+6}) p(w_{m+2}) \\ q(w_{m+5}) p(w_{m+3}) \\ \times q^3(w_{m+6}) p^2(w_{m+3}), & \text{if } w \in [w_{m+3}, w_{m+4}) \\ p(w_m) q^4(w_{m+5}) + q(w_{m+6}) p(w_{m+1}) q^3(w_{m+5}) \\ + q^2(w_{m+6}) p(w_{m+2}) q^2(w_{m+5}) \\ + q^3(w_{m+6}) p(w_{m+3}) q(w_{m+5}) + q^4(w_{m+6}) p(w_{m+4}), & \text{if } w \in [w_{m+5}, w_{m+6}) \\ q^5(w_{m+6}), \end{cases}$$

(5.17)

5.3 Equation Solved by the B-spline Basis Function   79

Table 5.9 $T_{m,5}(w)$ and its derivatives at connecting points.

	w_{m-2}	w_{m-1}	w_m	w_{m+1}	w_{m+2}	w_{m+3}
$T_{m,5}0$	$\left(\frac{h^2\tau^2+2p_1-2}{2\tau^2}\right)$		$r\left(\frac{h^2\tau^2-2p_1(h^2\tau^2+1)+2}{2\tau^2}\right)$	$r\left(\frac{h^2\tau^2-2p_1(h^2\tau^2+1)+2}{2\tau^2}\right)$	$r\left(\frac{h^2\tau^2+2p_1-2}{2\tau^2}\right)$	0
$T'_{m,5}0$	$-r\left(\frac{h\tau-p_2}{\tau}\right)$		$r\left(\frac{h\tau-3p_2+2p_1 h\tau}{\tau}\right)$	$-r\left(\frac{h\tau-3p_2+2p_1 h\tau}{\tau}\right)$	$r\left(\frac{h\tau-p_2}{\tau}\right)$	0
$T''_{m,5}0$	$r(1-p_1)$		$r(p_1-1)$	$r(p_1-1)$	$r(1-p_1)$	0
$T'''_{m,5}0$	$r\tau p_2$		$-3r\tau p_2$	$3r\tau p_2$	$-r\tau p_2$	0

5.2.10 Quartic trigonometric differential

A non-polynomial spline possesses smoothness depending on its degree over the interval.

The uniform partition of the interval [a, b] is considered by knots. The procedure of implementation of the Crank–Nicolson time integration and QTT B-spline basis finite element scheme is explained in this section. Let $[a, b]$ be a domain that is uniformly partitioned such that $a = w_0 < w_1 < \cdots < w_N = b$ and $h = w_{m+1} - w_m$, for $m = 0, 1, \ldots, M-1$. The QTT B-spline is defined as follows [25, 26]:

$$T_{m,5}(w) = r \begin{cases} \frac{\tau^2(-\hat{z}_{m-2})^2 + 2z_{m-2} - 2}{2\tau^2}, & x_{m-2} \leq x \leq x_{m-1}, \\ \frac{\tau^2\left(3h^2 + 6h\hat{z}_{m-2} + 2(-\hat{z}_{m-2})^2\right) + 2p_1\left(\tau^2(\hat{z}_{m-1})^2 - 2\right)}{2\tau^2}, & x_{m-1} \leq x \leq x_m, \\ \frac{\tau^2\left(13h^2 + 10h\hat{z}_{m-2} + 2(-\hat{z}_{m-2})^2\right)}{2\tau^2} & \\ \frac{+p_1\left(2\tau^2\left(11h^2 + 10h\hat{z}(m-2)\right) + 4p_1\tau^2(-\hat{z}_{m-2})^2 - 8p_1 + 6z_{m+1} - 4\right)}{2\tau^2}, & x_m \leq x \leq x_{m+1}, \\ \frac{\tau^2\left(23h^2 + 14h(\hat{z}_{m-2}) + 2(-\hat{z}_{m-2})^2\right) + 2p_1\left(\tau^2(\hat{z}_{m+2})^2\right)}{2\tau^2} & \\ \frac{-(2z_{m+1} + 6z_{m+2} - 4)}{2\tau^2}, & x_{m+1} \leq x \leq x_{m+2}, \\ \frac{\tau^2(\hat{z}_{m+3})^2 + 2z_{m+3} - 2}{2\tau^2}, & x_{m+2} \leq x \leq x_{m+3}, \\ 0, & \text{otherwise.} \end{cases}$$

(5.18)

where $\tau < \sqrt{\pi/h}$ is the tension parameter and $r = \frac{1}{2h^2(1-p_1)}$, $z_{m+j} = \cos(\tau(w_{m+j} - w))$, $\hat{z}_{m+j} = (w_{m+j} - w)$, $p_1 = \cos(\tau h)$, and $p_2 = \sin(\tau h)$ the value of the functions and its derivatives are tabulated in Table 5.9.

5.3 Equation Solved by the B-spline Basis Function

There are many partial differential equations in the literature, which are intensively investigated due to their wide range of applicability in engineering and other areas such as quantum mechanics, electromagnetic fields, fluid flow diffusion, etc. There are a variety of numerical techniques that can be used to solve partial differential equations such as differential quadrature method, finite element method, and collocation method. Some of these equations are 1D and 2D nonlinear stochastic quadratic integral equations, Fisher reaction-diffusion equations, Garden equation, nonlinear third-order Korteweg−de Vries (KdV) equations, equal width (EW) equations, modified Korteweg−de Vries (m-KdV) equations, Kuramoto−Sivashinsky equations, nonlinear modified Burger equations, time-fractional super-diffusion equations, fourth-order time fractional partial differential equations, modified regularized long wave (m-RLW) equations, nonlinear Schrodinger equations, linear and nonlinear BVP, singularly perturbed turning point BVP, and third-order singular by perturbed BVP.

5.4 Conclusion

In the present work, a review study has been done to discuss different B-spline basis function that are used in various well-known numerical methods that are widely used to solve nonlinear partial differential equations. These numerical methods are used to solve those differential equations whose solution is difficult to find analytically but are widely used in different fields of science and technology. The methods such as collcation method, differential quadrature method, finite element method can be applied to find out the solution of various problems having complexity in its structure. The present work can be considered as a direction for the researchers who are looking for basis functions and their formulas that can be utilize in the above mentioned approaches.

References

[1] Schoenberg I.J. (1946). Contribution to the problem of approximation of equidistant data by analytical functions, Quart. Appl. Math., 4, 45-99.
[2] Schumaker L.L. (1981). Spline functions, basic theory, Wiley.
[3] Cox M.G. (1972). The numerical evaluation of B-splines, Jour. Inst. Math. Apllic., 10,134-149.
[4] Boor C. de. (1978) A Practical Guide to Splines, Springer Verlag, New York.
[5] Kapoor, M., & Joshi, V. (2022). A novel technique for numerical approximation of 1D non-linear coupled Burgers' equation by using cubic Hyperbolic B-spline based Differential quadrature method. Turkish Journal of Computer and Mathematics Education, 13(2), 875-904.

[6] Kapoor, M., & Joshi, V. (2021). A new technique for numerical solution of 1D and 2D non-linear coupled Burgers' equations by using cubic Uniform Algebraic Trigonometric (UAT) tension B-spline based differential quadrature method. *Ain Shams Engineering Journal*, 12(4), 3947-3965.

[7] Görgülü, M., Dağ, İ., & Irk, D. (2015). Galerkin Method for the numerical solution of the RLW equation by using exponential B-splines.

[8] Ersoy, O., Dag, I., & Sahin, A. (2016). Numerical investigation of the solutions of Schrodinger equation with exponential cubic B-spline finite element method.

[9] Ersoy, O., & Dag, I. (2016). The Exponential Cubic B-Spline Collocation Method for the Kuramoto-Sivashinsky Equation. Filomat, 30(3), 853-861.

[10] Ersoy, O., & Dag, I. (2015). Numerical solutions of the reaction diffusion system by using exponential cubic B-spline collocation algorithms. Open Physics, 13(1).

[11] Ersoy, O., Dag, I., & Korkmaz, A. (2016). Solitary wave simulations of the Boussinesq Systems.

[12] Ersoy Hepson, O., & Dag, I. (2021). An exponential cubic B-spline algorithm for solving the nonlinear Coupled Burgers' equation. Computational Methods for Differential Equations, 9(4), 1109-1127.

[13] Hepson, O. E., Korkmaz, A., & İdiris, D. (2018). On the numerical solution of the Klein-Gordon equation by exponential cubic B-spline collocation method. Communications Faculty of Sciences University of Ankara Series A1 Mathematics and Statistics, 68(1), 412-421.

[14] Gorgulu, M., & Dag, I. (2018). Exponential B-splines Galerkin Method for the Numerical Solution of the Fisher's Equation. Iranian Journal of Science and Technology, Transactions A: Science, 42(4), 2189-2198.

[15] Hepson, O., Korkmaz, A., & Dag, I. (2020). Exponential B-spline collocation solutions to the Gardner equation. International Journal of Computer Mathematics, 97(4), 837-850.

[16] Kapoor, M., & Joshi, V. (2020). Solution of non-linear Fisher's reaction-diffusion equation by using Hyperbolic B-spline based differential quadrature method. In Journal of Physics: Conference Series, 1531 (1), 012064.

[17] Dag, I., & Hepson, O. (2021). Hyperbolic-trigonometric tension B-spline Galerkin approach for the solution of RLW equation. In AIP Conference Proceedings, 2334(1), 090005.

[18] Kaur, N., & Joshi, V. (2022). Numerical solution to the Gray-Scott Reaction-Diffusion equation using Hyperbolic B-spline. In Journal of Physics: Conference Series, 2267(1),012072.

[19] Arora, G., & Joshi, V. (2018). A computational approach using modified trigonometric cubic B-spline for numerical solution of Burgers' equation in one and two dimensions. Alexandria engineering journal, 57(2), 1087-1098.
[20] Kapoor, M., & Joshi, V. (2021). Numerical approximation of 1D and 2D non-linear Schrödinger equations by implementing modified cubic Hyperbolic B-spline based DQM. Partial Differential Equations in Applied Mathematics, 4, 100076.
[21] Kapoor, M., & Joshi, V. (2022). An Algorithm developed by implementing Modified Cubic Hyperbolic B-Spline based Differential Quadrature Method on non-linear Burgers' equation. In Journal of Physics: Conference Series 2267(1),012088.
[22] Kapoor, M., & Joshi, V. (2021). Numerical approximation of coupled 1D and 2D non-linear Burgers' equations by employing Modified Quartic Hyperbolic B-spline Differential Quadrature Method. *Int. J. Mech.*, *15*, 37-55.
[23] Kapoor, M., & Joshi, V. (2021). Numerical approximation of 1D and 2D reaction diffusion system with modified cubic UAH tension B-spline DQM. *J. Math. Comput. Sci.*, *11*(2), 1650-1667.
[24] Kapoor, M., & Joshi, V. (2021). A new technique for numerical solution of 1D and 2D non-linear coupled Burgers' equations by using cubic Uniform Algebraic Trigonometric (UAT) tension B-spline based differential quadrature method. *Ain Shams Engineering Journal*, 12(4), 3947-3965.
[25] Ersoy Hepson, Ö. (2021). A quartic trigonometric tension b-spline algorithm for nonlinear partial differential equation system. Engineering Computations, 38(5), 2293-2311.
[26] Ersoy Hepson, O., & Yigit, G. (2021). Quartic-trigonometric tension B-spline Galerkin method for the solution of the advection-diffusion equation. Computational and Applied Mathematics, 40(4), 1-15.
[27] Arora, G., & Joshi, V. (2019). Simulation of generalized nonlinear fourth order partial differential equation with quintic trigonometric differential quadrature method. Mathematical Models and Computer Simulations, 11(6), 1059-1083.
[28] Arora, G., & Joshi, V. (2016). Comparison of numerical solution of 1D hyperbolic telegraph equation using B-spline and trigonometric B-spline by differential quadrature method. Indian Journal of Science and Technology, 9(45), 1-8.

[29] Aji, N. T., & Joshi, V. (2020). The Numerical Study of Reaction-Diffusion Equation Systems Using Trigonometric Cubic B-Spline Differential Quadrature Method. European Journal of Molecular & Clinical Medicine, 7(07), 2020.

[30] Ersoy, O., & Dag, I. (2016). The numerical approach to the Fisher's equation via trigonometric cubic B-spline collocation method.

[31] Onarcan, A., Adar, N., & Dag, I. (2018). Trigonometric cubic B-spline collocation algorithm for numerical solutions of reaction–diffusion equation systems. Computational and Applied Mathematics, 37(5), 6848-6869.

[32] Aji, N. T., & Joshi, V. (2020). The Numerical Study Of Reaction-Diffusion Equation Systems Using Trigonometric Cubic B-Spline Differential Quadrature Method. European Journal of Molecular & Clinical Medicine, 7(07), 2020.

[33] Hepson, Ö. E. (2018). Numerical solutions of the Gardner equation via trigonometric quintic B-spline collocation method. Sakarya University Journal of Science, 22(6), 1576-1584.

[34] Korkmaz, A., Ersoy, O., & Dag, I. (2016). Trigonometric Cubic B-spline Collocation Method for Solitons of the Klein-Gordon Equation.

[35] Onarcan, A. T., Adar, N., & Dag, I. (2022). Pattern formation of Schnakenberg model using trigonometric quadratic B-spline functions. Pramana, 96(3), 1-9.

[36] Onarcan, A. T., Adar, N., & Dag, I. (2017). Numerical solutions of reaction-diffusion equation systems with trigonometric Quintic B-spline collocation algorithm.

[37] Joshi, V., & Kapoor, M. (2021). A Novel Technique for Numerical Approximation of 2-Dimensional Non-Linear Coupled Burgers' Equations using Uniform Algebraic Hyperbolic (UAH) Tension B-Spline based Differential Quadrature Method. Appl. Math, 15(2), 217-239.

[38] Kapoor, M., & Joshi, V. (2021) Numerical regime" Uniform Algebraic Hyperbolic Tension B-spline DQM" for the solution of Fisher's Reaction-Diffusion equation.

[39] Kapoor, M., & Joshi, V. (2021). A numerical regime for 1-D Burgers' equation using UAT tension B-spline differential quadrature method. International Journal for Computational Methods in Engineering Science and Mechanics, *22*(3), 181-192.

6

A Comparative Study: Modified Cubic B-spline-based DQM and Sixth-order CFDS for the Klein—Gordon Equation

G. Arora[1], B. K. Singh[2], Neetu Singh[3], and M. Gupta[2]

[1]Department of Mathematics, School of Chemical Engineering and Physical Sciences, Lovely Professional University, India
[2]Department of Mathematics, School of Physical and Decision Sciences, Babasaheb Bhimrao Ambedkar University, India
[3]Department of Applied Sciences and Humanities, KCNIT, India
E-mail: geetadma@gmail.com; brijeshks@bbau.ac.in; mukeshgupta.rs@bbau.ac.in; neetusinghmaths@gmail.com

Abstract

Numerical plays are of prime importance when the solution for modeled differential equation needs to be calculated. Numerous schemes are available for evaluating the approximate solution behaviors for these differential equations. But there is a need to compare the available approaches to know about the advantage of one method over another. This work provides a comparative study for the viability of two numerical schemes: modified cubic B-splines-based DQM (in brief, MCB-DQM) and sixth-order compact finite difference scheme (CFDS6). The above-mentioned schemes are tested for the Klein—Gordon (KG) equation (a generalization of the Schrodinger equation that was considered as the first formulated relativistic wave equation). The efficiency and accuracy of these methods are presented in terms of time-step, the accuracy, and the CPU time elapsed for the computation of the numerical approximations.

Keywords: differential quadrature method, sixth-order compact finite difference scheme, modified cubic B-splines, Klein—Gordon equation.

6.1 Introduction

This article is the study of the numerical solution obtained for KG equation with quadratic/cubic nonlinearity. Consider the KG equation, a relativistic wave equation, as follows:

$$u_{tt} + \nu u_{xx} + \lambda u + \gamma u^k = f(x,t), \quad x \in (x_L, x_R), \quad t > 0, \quad (6.1)$$

with initial/boundary conditions as per the concerned physical problem with ν, λ, γ being the known parameters

$$u(x, 0) = \phi(x), \quad u_t(x, 0) = \varphi(x), \quad (6.2)$$

$$u(x_L, t) = g_1(t), \quad u(x_R, t) = g_2(t), \quad (6.3)$$

where $u(x,t)$ is the wave displacement and $f(x,t)$ signifies the source term. The KG equation is solved with quadratic ($\gamma \neq 0$, $k = 2$) and cubic ($\gamma \neq 0$, $k = 3$) nonlinearity. This equation has applications and exists in the study of various scientific phenomena such that it is considered as a quantum wave equation and has been derived as a part of the well-known Schrodinger equation. It also exists in the study of solid state physics and electromagnetic interaction in quantum field theory [1] and in nonlinear optics.

The study of the behavior of the time-dependent partial differential equations (PDEs) plays a significant role to understand the various phenomena that occurred in terms of these equations or the systems of these equations. In the past, various techniques have been introduced to complete this purpose, and some of the well-known techniques are collocation techniques with different base functions[2, 3], homotopy perturbation method [4, 5], spectral method [6], pseudo-spectral methods [7], B-spline collocation method [8, 9], radial basis function (RBF) [10], lattice Boltzmann method [11], homotopy perturbation transform method [12], decomposition method [13], tanh and the sine−cosine methods [14], variation iteration method [15], J-transform-based optimal HAM and variational iteration VIT [16, 17] fractional differential transform method [18–23], etc.

The solution of this equation has also been studied in many articles for the soliton-solutions that appear with their interaction in a collision-less plasma [13, 14]. Several numerical schemes to solve KG equation (6.1) have been developed by many researchers. Some of them are as follows: homotopy perturbation method [4], spectral method [6], pseudo-spectral methods [7], B-spline collocation method [8, 9], radial basis function (RBF) [10], lattice Boltzmann method [11], decomposition method [13], tanh and the sine−cosine methods [14], and many more.

In recent years, various (non) linear differential equations appearing in one-, two-, or higher-dimensional has been successfully by the application of the DQM. In this approach, to calculate the weighting coefficients, different forms of test functions can be implemented, which include the use of exponential spline functions, Laguerre polynomials, Lagrange polynomials, barycentric Lagrange interpolation basis function, radial basis function, and many more; see [24–34] and the references therein.

The CFDS6 is a well-known algorithm to get the approximate behavior of various types of PDEs [35, 36]. It has been implemented to solve hyperbolic equations [37], Euler, and Navier–Stokes equations [38]. The Schrödinger equation has been solved by a compact split-step FDS [39], and reaction–diffusion equations have been solved by a high-order CFDS [40]. The CHDS6 has been utilized to get an approximate behavior of Helmholtz equation [41], integro-differential equations [42], Fisher's equation [43], and Burgers' equation [44].

This article is concerned with the comparison of numerical solution for the KG equation obtained by two schemes: MCB-DQM [29] and CFDS6 [36, 42, 43]. The transformation $u_t = c$ with eqn (6.1) is converted into a system of PDEs, and, thereafter, on putting spatial derivative values obtained by either MCB-DQM or CFDS6, the resulting system of PDEs is converted into the system of ODEs of order 1 which is solved via the SSP-RK43 scheme [47].

In MCB-DQM, a modified cubic-B-spline basis function is utilized in DQM to find the derivative approximations, while in CFDS6, finite difference formula with some constants based on the required accuracy is utilized for second-order spatial derivative. The comparative numerical study is then performed on the three numerical test problems to establish the methods. The comparison of L_2 and L_∞ errors is obtained by the following schemes: MCB-DQM and CFDS6 are done with the results available in the literature.

In the next section, both methods will be discussed with the detailed methodology and implementation procedure described in Section 6.3. It is followed by the three numerical test examples taken with different parameter values in Section 6.4. After the presentation of the obtained results in the form of tables and figures, the article is concluded in Section 6.5.

6.2 Methodology

The first step to get an approximate solution behavior to KG equation is the discretization of the considered finite interval. To find the values of spatial derivatives at each grid, either MCB-DQM or CFDS6 is utilized in the present work. The spatial interval is distributed uniformly into N grid (node) points, i.e., $x_L = x_1 < x_2 < \cdots < x_{N-1} < x_N = x_R$ with $h = x_{j+1} - x_j = \frac{(x_R - x_L)}{N-1}$. Following are the procedural steps for calculating the spatial derivatives by both the considered approaches

6.2.1 MCB-DQM

The fairly accurate approximations of spatial derivatives of order one/two at x_i, $i = 1, \ldots, N$ are given by
$u_x(x_i, t) = \sum_{j=1}^{N} \alpha_{ij} u(x_j, t)$ and $u_{xx}(x_i, t) = \sum_{j=1}^{N} \beta_{ij} u(x_j, t)$, where α_{ij} and β_{ij} are the weighting coefficients that are to be obtained using the chosen basis functions and hence provide the derivatives [46].

In the present work, modified cubic B-spline basis function φ_i, $i = 1, 2, \ldots, N$, is utilized to obtain the values of α_{ij} and β_{ij} [29]:

$$\left.\begin{aligned}
&\varphi_1(x) = \phi_1(x) + 2\phi_0(x), \quad \varphi_2(x) = \phi_2(x) - \phi_0(x) \\
&\vdots \qquad \vdots \\
&\varphi_j(x) = \phi_j(x), \quad j = 3, 4, \ldots, N-2 \\
&\vdots \qquad \vdots \\
&\varphi_{N-1}(x) = \phi_{N-1}(x) - \phi_N(x), \quad \varphi_N(x) = \phi_N(x) + 2\phi_N(x)
\end{aligned}\right\}, \quad (6.4)$$

where ϕ_i, $i = 1, 2, \ldots, N$, are the classical cubic B-spline functions that provide the values of φ_j' and φ_j'' at different knot points, as shown in [29]. The procedure for finding these weighting coefficients [29] is reported in the following section.

6.2.1.1 The weighting coefficients

The spatial derivative of order 1 at grid x_i, $i = 1, 2, \ldots, N$ is given by

$$\varphi_k'(x_i) = \sum_{j=1}^{N} \alpha_{ij} \varphi(x_j), \quad k = 1, 2, \ldots, N. \quad (6.5)$$

After replacing the values of $\phi_k'(x_i)$'s and $\phi_k''(x_i)$'s, eqn (6.5) results in a tridiagonal system of linear equations as follows:

$$X\vec{\alpha}[i] = Y[i], \quad i = 1, 2, \ldots, N,$$

where X denotes the coefficient matrix with the main diagonal entries as 4, and the super and sub-diagonal entries as 1 with changes in the first and the last rows that can be defined as follows:

$$X(i, i) = 4, \quad X(i, i-1) = X(i+1, i) = 1, \quad \forall \ i = 1, 2, \ldots, N$$
$$X(1, 1) = X(N, N) = 6; \quad X(2, 1) = X(N-1, N) = 0,$$

and $\vec{\alpha}[i]$ is used for the weighting-coefficient associated with grid point x_i, i.e., $\vec{\alpha}[i] = [\alpha_{i1}, \alpha_{i2}, \ldots, \alpha_{iN}]^T$, and $\vec{Y}[i] = [y_{i1}, y_{i2}, \ldots, y_{iN}]^T$ is evaluated as

$$\vec{Y}[1] = \begin{bmatrix} -6/h \\ 6/h \\ 0 \\ \vdots \\ 0 \end{bmatrix}, \vec{Y}[2] = \begin{bmatrix} -3/h \\ 0 \\ 3/h \\ \vdots \\ 0 \end{bmatrix}, \ldots,$$

$$\vec{Y}[N-1] = \begin{bmatrix} 0 \\ \vdots \\ -3/h \\ 0 \\ 3/h \end{bmatrix}, \vec{Y}[N] = \begin{bmatrix} 0 \\ \vdots \\ 0 \\ -6/h \\ 6/h \end{bmatrix}.$$

The resulting tridiagonal system of linear equations on solving provides the weighting coefficients $\alpha_{i1}, \alpha_{i2}, \ldots, \alpha_{iN}$ for $i = 1, 2, \ldots, N$, and that for finding the second-order spatial derivative is evaluated via the following:

$$\beta_{ij} = \alpha_{ij}\left(\alpha_{ii} - \frac{1}{x_i - x_j}\right), \qquad i \neq j,$$

$$\beta_{ii} = \sum_{i=1,\, i\neq j}^{N} \beta_{ij}, \qquad \text{Else.}$$

6.2.2 CFDS6

The approach of CFDS6 [36, 42, 43] can be considered as an advancement in the finite difference method in which the derivatives are approximated via the finite difference formula with some constants based on the required accuracy. As per the scheme, the spatial derivative of order 2 at each internal collocation point can be evaluated from [36], as follows:

$$\theta u''_{i-1} + u''_i + \theta u''_{i+1} = b\frac{u_{i+2} - 2u_i + u_{i-2}}{4h^2} + a\frac{u_{i+1} - 2u_i + u_{i-1}}{h^2}, \quad (6.6)$$

and for $a = \frac{4}{3}(1-\theta)$, $b = \frac{1}{3}(-1+10\theta)$, eqn (6.6) is reduced into a θ-family of fourth-order tridiagonal schemes. The CFDS6 for the second-order derivative is evaluated by fixing $\theta = 1/10$ in eqn (6.6), read as

$$u''_{i-1} + 10u''_i + u''_{i+1} = \frac{12}{h^2}(u_{i+1} - 2u_i + u_{i-1}).$$

For the nodes x_1 and x_N on the boundary, the approximation to the derivative of order 2 can be derived by using the Taylor series expansion about the nodes [42, 43]. Detailed derivations for the derivatives can be read from available work [40, 42, 43, 44].

The derived formulae [43] for second-order derivative at the nodes x_1 and x_N, respectively, are

$$10u''_1 + u''_2 = \frac{12}{h^2}\left(\frac{115}{36}u_1 - \frac{1555}{144}u_2 + \frac{89}{6}u_3 - \frac{773}{72}u_4 + \frac{151}{36}u_5 - \frac{11}{16}u_6\right),$$

$$u''_{N-1} + 10u''_N = \frac{12}{h^2}\left(-\frac{11}{16}u_{N-5} + \frac{151}{36}u_{N-4} - \frac{773}{72}u_{N-3} + \frac{89}{6}u_{N-2}\right.$$
$$\left. - \frac{1555}{144}u_{N-1} + \frac{115}{36}u_N\right).$$

The above scheme in matrix form can be read as

$$\begin{bmatrix} 10 & 1 & & & \\ 1 & 10 & 1 & & \\ & \ddots & \ddots & \ddots & \\ & & 1 & 10 & 1 \\ & & & 1 & 10 \end{bmatrix} \begin{bmatrix} u''_1 \\ u''_2 \\ \vdots \\ u''_{N-1} \\ u''_N \end{bmatrix} = \begin{bmatrix} b_1(u) \\ b_2(u) \\ \vdots \\ b_{N-1}(u) \\ b_N(u) \end{bmatrix},$$

with

$$b_1(u) = \frac{12}{h^2}\left(\frac{115}{36}u_1 - \frac{1555}{144}u_2 + \frac{89}{6}u_3 - \frac{773}{72}u_4 + \frac{151}{36}u_5 - \frac{11}{16}u_6\right);$$
$$b_i(u) = \frac{12}{h^2}(u_{i-1} - 2u_i + u_{i+1}), \quad i = 2, 3, ..., N-1;$$
$$b_N(u) = \frac{12}{h^2}\left(-\frac{11}{16}u_{N-5} + \frac{151}{36}u_{N-4} - \frac{773}{72}u_{N-3} + \frac{89}{6}u_{N-2}\right.$$
$$\left. - \frac{1555}{144}u_{N-1} + \frac{115}{36}u_N\right).$$

The resulting tridiagonal matrix system is solved via the "Thomas algorithm" which provides second-order spatial derivatives at each node.

6.3 Implementation of the Method

First, using the transformation $u_t = c$ to convert eqn (6.1) into the system of first-order PDEs results in

$$u_t = c$$
$$c_t + \nu u_{xx} + \lambda u + \gamma u^k = f(x,t). \tag{6.7}$$

After using the values of the second derivative of either MCB-DQM or CFDS6, system (6.7) of PDEs is reduced to a coupled system of ODEs of first order:

$$\frac{du_i}{dt} = c_i, \tag{6.8}$$
$$\frac{dc_i}{dt} = Z(u_i, c_i, f_i), \quad i = 1, 2, ..., N,$$

where Z represents a nonlinear differential operator. This obtained system of eqn (6.8) is hence solved implementing the BCs as defined in eqn (6.3) and the ICs

$$u(x_i, 0) = g_1(x_i), \quad c(x_i, 0) = g_2(x_i), \quad i = 1, ..., N.$$

The obtained system of ODEs (6.8) can be solved by any numerical approach, but, in this work, the SSP-RK43 scheme is used to find the solutions [45]. Using the solution at the lower time level approach, the solution at the required time level can be obtained recursively.

6.4 Results and Discussion

To gain an insight for the accuracy/efficiency of the methods, three different examples of KG equations are analyzed for MCB-DQM and CFDS6.

Example 1

KG equation (6.1) with cubic nonlinearity ($\gamma \neq 0$, $k = 3$) is solved for the parameters $\alpha = -1, \beta = 3/4, \gamma = -3/2$, and $f = 0$

$$u_{tt}(x,t) = \frac{3}{2}u^3(x,t) + u_{xx}(x,t) - \frac{3}{4}u(x,t).$$

For BCs/ICs condition taken from the exact solution: $u(x,t) = -\sec h(x + t/2)$.

The enumerated solutions in [0, 1] with $h = 0.02$ and time-step $k = 0.0001$ are compared in terms of absolute errors via MCB-DQM/CFDS6

Table 6.1 Absolute errors via MCB-DQM and CFDS6 for Example 1 at time t, $t \leq 0.4$.

x/T	0.1		0.2		0.3		0.4	
	MCB-DQM	CFDS6	MCB-DQM	CFDS6	MCB-DQM	CFDS6	MCB-DQM	CFDS6
0	2.46E-06	2.49E-06	4.71E-06	4.96E-06	6.57E-06	7.36E-06	7.89E-06	9.67E-06
0.1	6.18E-06	3.76E-08	1.51E-05	1.22E-06	1.64E-05	2.54E-06	1.72E-05	3.80E-06
0.2	5.78E-10	1.06E-10	6.09E-06	4.14E-08	1.56E-05	1.23E-06	1.70E-05	2.58E-06
0.3	1.41E-11	3.20E-11	9.72E-10	8.32E-11	6.05E-06	4.39E-08	1.61E-05	1.24E-06
0.4	1.65E-11	3.70E-11	7.24E-11	1.63E-10	2.25E-10	3.64E-10	6.02E-06	4.55E-08
0.5	3.75E-11	8.42E-11	1.47E-10	3.31E-10	3.09E-10	6.94E-10	1.08E-09	2.07E-09
0.6	4.67E-11	1.05E-10	1.79E-10	4.01E-10	4.53E-10	8.67E-10	7.30E-07	4.68E-06
0.7	4.57E-11	1.03E-10	2.97E-10	5.10E-11	7.33E-07	4.64E-06	2.38E-06	1.32E-05
0.8	7.83E-11	4.21E-10	7.39E-07	4.59E-06	2.33E-06	1.36E-05	3.03E-06	1.45E-05
0.9	7.49E-07	4.53E-06	2.28E-06	1.38E-05	2.98E-06	1.42E-05	3.67E-06	1.09E-05
1	3.84E-07	2.44E-05	9.66E-07	2.40E-05	1.67E-06	2.35E-05	2.44E-06	2.30E-05

Table 6.2 Thousand times of the absolute errors via CFDS6, MCB-DQM, and TMM [47] for Example 1 at $t = 0.5$.

Schemes	x = 0	x = 0.1	x = 0.2	x = 0.3	x = 0.4	x = 0.5
CFDS6	1.19E-02	4.99E-03	3.89E-03	2.61E-03	1.25E-03	4.77E-03
MCB-DQM	8.61E-03	1.76E-02	1.81E-02	1.75E-02	1.64E-02	5.27E-03
TMM [47]	0	4.14E-01	3.39E-01	3.54E-01	1.19E-01	7.83E-01
	x = 0.6	x = 0.7	x = 0.8	x = 0.9	x = 1	
CFDS6	1.28E-02	1.47E-02	1.13E-02	1.07E-02	2.25E-02	
MCB-DQM	2.41E-03	3.07E-03	3.73E-03	4.34E-03	3.22E-03	
TMM [47]	6.13E-01	1.41E-01	7.68E-01	2.71E-01	8.41E-01	

at time $t \leq 0.4$ and are presented in Table 6.1, while the comparison of evaluated absolute errors at $t = 0.5$ with TMM [47] is reported in Table 6.2. L_2 and L_∞ errors are compared in Table 6.3 for MCB-DQM/CFDS6 at $t \leq 0.5$. The 2D plots of the absolute errors via MCB-DQM and CFDS6 at $t \leq 0.5$ are depicted in Figure 6.1, and Figure 6.2 depicts the behavior the referred problem. The comparisons are made at various time levels for the absolute L_2 and L_∞ errors. These findings confirm that both schemes are giving equivalent results and better than TMM [47].

Table 6.3 L_∞ and L_2 errors for Example 1 via MCB-DQM and CFDS6 at time t, $t \leq 0.5$.

Schemes	Errors	$t = 0.1$	$t = 0.2$	$t = 0.3$	$t = 0.4$	$t = 0.5$
MCB-DQM	$L_\infty \times 10^3$	1.49E-02	1.57E-02	1.66E-02	1.73E-02	1.81E-02
CFDS6		1.54E-02	1.56E-02	1.57E-02	1.57E-02	1.58E-02
MCB-DQM	$L_2 \times 10^3$	4.15E-03	6.50E-03	8.48E-03	1.03E-02	1.20E-02
CFDS6		3.54E-03	5.38E-03	6.75E-03	7.86E-03	8.91E-03

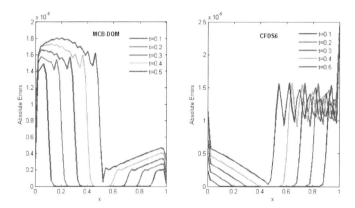

Figure 6.1 Absolute errors for Example 1 at $t \leq 0.5$ obtained by MCB-DQM and CFDS6.

Example 2

KG equation (6.1) having quadratic-nonlinearity

$$u_{tt}(x,t) = u_{xx}(x,t) - u^2(x,t) + 6xt(x^2 - t^2) + x^6 t^6,$$

with BCs taken from the exact solution [9] of the equation $u(x,t) = x^3 t^3$.

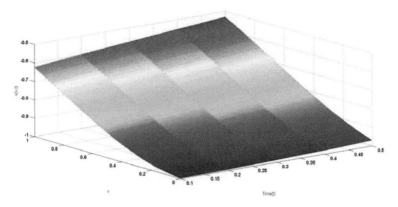

Figure 6.2 Physical behavior of the solution of Example 1 at $t \leq 0.5$.

The enumerated behaviors, in terms of L_2 and L_∞ errors and computational (CPU) time, over domain $[0, 1]$ with $h = 1/60$; $k = 0.0001$, are compared for 0MCB-DQM and CFDS6 at time $t \leq 5$, in Table 6.4. In Table 6.5, the comparison of L_2 and L_∞ errors and the computational time for MCB-DQM, CFDS6, and RBF [10] are done for $h = 0.02$; $k = 0.0001$. The absolute errors by MCB-DQM and CFDS6 at different time levels $t \leq 5$ are plotted in Figure 6.3, while the behavior is depicted graphically in Figure 6.4.

From the comparison, it is evident that the errors in the numerical solution obtained by the MCB-DQM and CFDS6 are comparable and are better than the results obtained by RBF [10]. Also the computational time in CFDS6 is less compared to CPU time in MCB-DQM.

Table 6.4 Comparison for Example 2 taking $h = 1/60$

t	MCB-DQM			CFDS6		
	L_2	L_∞	Time	L_2	L_∞	Time
1	4.4428E-05	9.3022E-05	1.70	6.50E-05	1.45E-04	0.50
2	8.1610E-05	1.5901E-04	3.42	2.80E-04	5.72E-04	1.00
3	6.2802E-05	1.8398E-04	5.06	3.95E-04	1.21E-03	1.62
4	2.7142E-04	9.5947E-04	6.77	5.10E-04	2.02E-03	1.96
5	5.6348E-04	2.3355E-03	8.41	6.27E-04	2.93E-03	2.49

6.4 Results and Discussion

Table 6.5 Comparison in Example 2 for MCB-DQM, CFDS6, and RBF [10] for $h = 0.02$.

T	MCB-DQM			CFDS6			RBF [10]		
	L_2	L_∞	Time	L_2	L_∞	Time	L_2	L_∞	Time
1	4.4428E-05	6.9788E-05	1.21	6.47E-05	1.44E-04	0.405	5.4998E-05	1.1012E-05	6
2	8.1610E-05	5.9226E-05	2.40	2.79E-04	5.66E-04	0.811	1.1522E-03	1.6496E-04	14
3	6.2802E-05	7.2453E-04	3.64	3.91E-04	1.19E-03	1.201	3.2588E-03	5.9728E-04	25
4	2.7142E-04	2.1151E-03	4.95	5.02E-04	1.95E-03	1.606	9.8191E-03	1.8264E-03	37
5	5.6348E-04	4.4562E-03	6.04	6.11E-04	2.79E-03	1.996	1.9139E-02	3.6915E-03	52

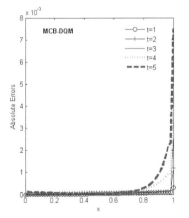 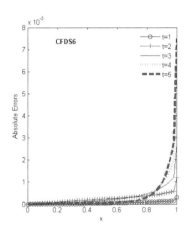

Figure 6.3 Absolute errors for Example 2 at $t \leq 5$ obtained by MCB-DQM and CFDS6.

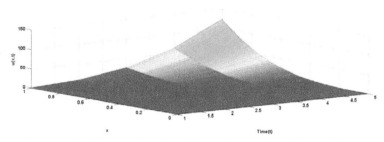

Figure 6.4 Physical behavior of the solution of Example 2 at $t \leq 5$.

Example 3. Take KG equation (6.1) having quadratic-nonlinearity

$$u_{tt}(x,t) = u_{xx}(x,t) - u^2(x,t) - x\cos t + x^2 \cos^2 t,$$

with the initial condition $u(x,0) = x$ and $u_t(x,0) = 0$. The exact behavior [10] is $u(x,t) = x\cos t$.

The BCs are taken from the exact solution. The numerical solutions are obtained over the domain $[-1, 1]$ taking $h = 0.1$ and the time-step $k = 0.0001$. The comparison of L_2 and L_∞ errors obtained by MCB-DQM and CFDS6 with the errors due to RBF [10] at different time levels $t \leq 10$ is reported in Table 6.6 along with the computational (CPU) time. Also L_2 and L_∞ errors are compared with TMM [47] at different time levels $t \leq 1$ and are reported in Table 6.7. It is evident that the results obtained by our method are better than that in [10, 44]. The physical behavior of the solution is depicted in Figure 6.5.

Table 6.6 Comparison of errors and computational time in Example 3 for MCB-DQM, CFDS6, and RBF [10].

T	MCB-DQM			CFDS6			RBF [10]		
	L_2	L_∞	CPU	L_2	L_∞	CPU	L_2	L_∞	CPU
1	3.17E-05	3.82E-05	0.39	3.16E-05	3.77E-05	0.296	6.54E-05	1.25E-05	5
3	6.05E-06	6.98E-06	1.17	6.02E-06	7.55E-06	0.889	1.17E-04	1.56E-05	22
5	3.64E-05	4.41E-05	1.95	3.64E-05	4.50E-05	1.466	2.20E-04	3.38E-05	49
7	2.56E-05	3.05E-05	2.73	2.55E-05	3.00E-05	2.137	2.59E-04	3.78E-05	83
10	2.06E-05	2.39E-05	3.90	2.14E-05	2.66E-05	2.948	7.99E-05	1.31E-05	150

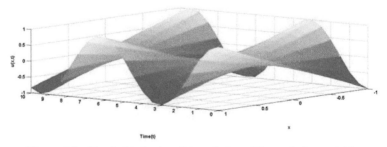

Figure 6.5 Physical behavior of the solution of Example 3 at $t \leq 10$.

Table 6.7 Comparison in Example 3 for MCB-DQM, CFDS6, and TMM [47].

Schemes	$t=0.1$	$t=0.2$	$t=0.3$	$t=0.4$	$t=0.5$	$t=0.6$	$t=0.7$	$t=0.8$	$t=0.9$	$t=1$
					L_2					
MCB-DQM	8.34E-7	4.73E-6	9.95E-6	1.48E-5	1.94E-5	2.37E-5	2.79E-5	3.17E-5	3.51E-5	3.82E-5
CFD6	6.57E-7	4.32E-6	1.03E-5	1.54E-5	1.89E-5	2.31E-5	2.83E-5	3.22E-5	3.47E-5	3.77E-5
TMM[47]	4.96E-8	5.65E-7	1.85E-6	3.28E-6	3.84E-6	4.62E-6	1.41E-5	5.69E-5	1.86E-4	2.23E-4
					L_∞					
MCB-DQM	3.68E-7	2.12E-6	4.84E-6	8.04E-6	1.16E-5	1.54E-5	1.93E-5	2.34E-5	2.75E-5	3.17E-5
CFD6	2.89E-7	1.92E-6	4.91E-6	8.31E-6	1.16E-5	1.52E-5	1.93E-5	2.35E-5	2.76E-5	3.16E-5
TMM[47]	9.61E-6	5.49E-6	1.19E-5	1.58E-5	1.47E-5	1.47E-5	3.94E-5	1.40E-4	4.05E-4	9.80E-4

6.5 Conclusion

In this chapter, KG equation is enumerated by two schemes: MCB-DQM and CFDS6. By using both the schemes, the value of second-order derivative is obtained and the PDE is reduced to a system of ODE. The formulated system of ODEs of order 1 is enumerated via the SSP-RK43 scheme. The accuracy/efficiency of the methods is measured by taking three examples and comparing the computational (CPU) time and L_2 and L_∞ errors with that given in the literature. It can be concluded that both the schemes give comparable results, but the computational time in CFDS6 is less as compared to MCB-DQM while the errors obtained by both the schemes are comparable. Both of these schemes are easy to implement to getting comparatively more accurate solution to the linear/nonlinear PDEs.

References

[1] A. M. Wazwaz, New travelling wave solutions to the Boussinesq and the Klein-Gordon equations, Commu. Nonlinear Sci. Numer. Simulat. 13 (2008) 889-901.

[2] B.K. Singh, M. Gupta, A new efficient fourth order collocation scheme for solving Burgers' equation. Appl. Math. Comput. **399** (15), 126011 (2021). https://doi.org/10.1016/j.amc.2021.126011

[3] B.K. Singh, M. Gupta, Trigonometric tension B-spline collocation approximations for time fractional Burgers' equation, Journal of Ocean Engineering and Science (2022), https://doi.org/10.1016/j.joes.2022.03.023.

[4] Z. Odibat and S. Momani, A reliable treatment of homotopy perturbation method for Klein-Gordon equations, Physics Letters A. 365 (5-6) (2007) 351-357.

[5] B.K. Singh, P. Kumar, V. Kumar, Homotopy perturbation method for solving time fractional coupled viscous Burgers' equation in (2+1) and (3+1) dimensions. Int. J. Appl. Comput. Math.**4**(38), (2018)

[6] B.Y. Guo, X. Li and L. Vazquez, A Legendre spectral method for solving the nonlinear Klein-Gordon equation, Math. Appl. Comput. 15 (1) (1996) 19-36.

[7] X. Li and B.Y. Guo, A Legendre spectral method for solving nonlinear Klein-Gordon equation, J. Comput. Math. 15 (2) (1997) 105-126.

[8] J. Rashidinia, F. Esfahani and S. Jamalzadeh, B-spline collocation approach for solution of Klein-Gordon equation, Int. J. Math. Modelling Comput. 3 (1) (2013) 25-33.

[9] J. Rashidinia, M. Ghasemi and R. Jalilian, Numerical solution of the nonlinear Klein-Gordon equation, J. Comput. Appl. Math. 233 (2010) 1866-1878.

[10] M. Dehghan and A. Shokri, Numerical solution of the nonlinear Klein-Gordon equation using radial basis functions, J. Comput. Appl. Math. 230 (2009) 400-410.

[11] Q. Li, Z. Ji, Z. Zheng and H. Liu, Numerical Solution of Nonlinear Klein-Gordon Equation Using Lattice Boltzmann Method, Applied Mathematics 2 (2011) 1479-1485.

[12] B.K. Singh, P. Kumar, Homotopy perturbation transform method for solving fractional partial differential equations with proportional delay. SeMA J. **75**, 111–125 (2018)

[13] S. M. El-Sayed, The decomposition method for studying the Klein-Gordon equation, Chaos Solitons Fractals 18 (2003) 1025-1030.

[14] A. M. Wazwaz, The tanh and the sine-cosine methods for compact and noncompact solutions of the nonlinear Klein-Gordon equation, Appl. Math. Comput. 167 (2005) 1179-1195.

[15] B.K. Singh, P. Kumar, Fractional variational iteration method for solving fractional partial differential equations with proportional delay. Int. J. Differ. Eqn. **88** (8), 1–11 (2017)

[16] B.K. Singh, A. Kumar, M. Gupta, Efficient New Approximations for Space-Time Fractional Multi-dimensional Telegraph Equation, Int. J. Appl. Comput. Math (2022) 8:218.

[17] B.K. Singh, A. Kumar, New approximate series solutions of conformable time–space fractional Fokker–Planck Equation via two efficacious techniques, Partial Differential Equations in Applied Mathematics 6, (2022) 100451.

[18] B.K. Singh, P. Kumar, Extended fractional reduced differential transform for solving fractional partial differential equations with proportional delay. Int. J. Appl. Comput. Math. **3**(1), 631–649 (2017)

[19] B.K. Singh, S. Agrawal, A new approximation of conformable time fractional partial differential equations with proportional delay. Appl. Numer. Math. **157**, 419–433 (2020)

[20] B.K. Singh, M. Gupta, A comparative study of analytical solutions of space-time fractional hyperbolic like equations with two reliable methods. Arab. J. Basic Appl. Sci. **26** (1), 41–57 (2019)

[21] B.K. Singh, V.K. Srivastava, Approximate series solution of multi-dimensional, time fractional-order (heat-like) diffusion equations using FRDTM. R. Soc. Open Sci. **2**(5), 140511 (2015)

[22] B.K. Singh, P. Kumar, FRDTM for numerical simulation of multi-dimensional, time-fractional model of Navier-Stokes equation. Ain Shams Eng. J. **9**(4), 827–834 (2018)

[23] B.K. Singh, S. Agrawal, Study of time fractional proportional delayed multi-pantograph system and integro-differential equations, Math. Meth. Appli. Sci. 2022; 45:8305–8328, https://doi.org/10.1002/mma.8335

[24] J. R. Quan and C.T. Chang, New insights in solving distributed system equations by the quadrature methods-I, Comput. Chem. Eng. 13 (1989) 779-788.

[25] J. R. Quan and C. T. Chang, New insights in solving distributed system equations by the quadrature methods-II, Comput. Chem. Eng. 13 (1989) 1017-1024.

[26] C. Shu and B. E. Richards, Application of generalized differential quadrature to solve two dimensional incompressible Navier-Stokes equations, Int. J. Numer. Meth. Fluids. 15 (1992) 791-798.

[27] Abdul-Sattar Jaber Al-Saif, Afrah Abdul-Imam Almazinyb, Using Laguerre Polynomials as a Basis for A new Differential Quadrature Methodology to Solve Magneto-Hydrodynamic (MHD) Fourth-Grade Fluid Flow, Int. J. Nonlinear Anal. Appl., 12, (2021) 83-96.

[28] Varun Joshi, Mamta Kapoor, Nitin Bhardwaj, Mehedi Masud,2and Jehad F. Al-Amri, Numerical Approximation of One-andTwo-Dimensional Coupled Nonlinear Schrodinger Equation by Implementing Barycentric Lagrange Interpolation Polynomial DQM, Mathematical Problems in Engineering (2021).

[29] G. Arora and B. K. Singh, Numerical solution of Burgers' equation with modified cubic B-spline differential quadrature method, Appl. Math. Comput. 224 (2013) 166-177.

[30] A. Korkmaz, Shock wave simulations using sinc differential quadrature method, Int. J. Comput. Aided Eng. Software 28 (6) (2011) 654-674.

[31] B.K. Singh, J.P. Shukla, M. Gupta, Study of one dimensional hyperbolic telegraph equation via a hybrid cubic B-spline differential quadrature method. Int. J. Appl. Comput. Math. **7**(1), 14 (2021).

[32] B.K. Singh, P Kumar, An algorithm based on a new DQM with modified extended cubic B-splines for numerical study of two dimensional hyperbolic telegraph equations. Alex. Eng. J. **57**(1), 175–191 (2018)

[33] B.K. Singh, Anovel approach for numeric study of 2D biological population model. Cogent Math. **3**, 1261527 (2016). https://doi.org/10.1080/23311835.2016.1261527

[34] B.K. Singh, P. Kumar, A novel approach for numerical computation of Burgers' equation in (1 +1) and (2 + 1) dimensions. Alex. Eng. J. **55**(4), 3331–3344 (2016)

[35] I. Christie, Upwind compact finite difference schemes, J. Comput. Phys. (59) 3 (1985) 353-368.

[36] S. K. Lele, Compact finite difference schemes with spectral-like resolution, J. Comput. Phys. 103(1) (1992) 16-42.

[37] H. J. Wirz, F. D. Schutter and A. Turi, An implicit, compact, finite difference method to solve hyperbolic equations, Math. Comput. Simulat. 19(4) (1977) 241-261.

[38] M. E. Rose, Compact finite difference schemes for the Euler and Navier-Stokes equations, J. Comput. Phys. 49 (3) (1983) 420-442.

[39] M. Dehghan and A. Taleei, A compact split-step finite difference method for solving the nonlinear Schrödinger equations with constant and variable coefficients, Comp. Phys. Commun. 181 (1) (2010) 43-51.

[40] Y. M. Wang and H. B. Zhang, Higher-order compact finite difference method for systems of reaction diffusion equations, J. Comput. Appl. Math. 233 (2009) 502-518

[41] G. Sutmann, Compact finite difference schemes of sixth order for the Helmholtz equation, J. Comput. Appl. Math. 203 (1) (2007) 15-31.

[42] J. Zhao and R. M. Corless, Compact finite difference method for integro-differential equations, Appl. Math. Comput. 177(1) (2006) 271-288.

[43] M. Bastani and D. K. Salkuyeh, A highly accurate method to solve Fisher's equation, Pramana – J. Phys. Indian Academy of Sciences, 78(3) (2012) 335-346.

[44] M. Sari and G. Gürarslan, A sixth-order compact finite difference scheme to the numerical solutions of Burgers' equation, Appl. Math. Comput. 208(2) (2009) 475-483

[45] J. R. Spiteri and S. J. Ruuth, A new class of optimal high-order strong stability-preserving time-stepping schemes, SIAM J. Numer. Anal. 40(2) (2002) 469-491.

[46] R. Bellman, B. G. Kashef and J. Casti, Differential quadrature: a technique for the rapid solution of nonlinear differential equations, J. Comput. Phy. 10 (1972) 40-52.

[47] B. Bulbul and M. Sezer, A new approach to numerical solution of nonlinear Klein-Gordon equation, Mathematical Problems in Engineering 2013, Article ID 869749, 7 pages, http://dx.doi.org/10.1155/2013/869749.

7

Sumudu ADM on Time-fractional 2D Coupled Burgers' Equation: An Analytical Aspect

Mamta Kapoor

Department of Mathematics, Lovely Professional University, Phagwara, Punjab, India-144411
E-mail: mamtakapoor.78@yahoo.com

Abstract

The analytical solution of 2D nonlinear fractional coupled Burgers' equation is the topic of the present chapter. In the present study, a new method called Sumudu ADM is created and applied to the 2D nonlinear coupled time-fractional Burgers' problem. An elegant combination of Sumudu transform and ADM is Sumudu ADM. The obtained findings are closely related to the precise answers.

Keywords: Sumudu transform, adomian decomposition method, Sumudu ADM, 2D nonlinear time-fractional coupled Burgers' equation.

7.1 Introduction

In recent years, a lot of research has been done regarding numerical and analytical solutions of the physical prototypes of some well-known problems. Some of the work in this regard are as per literature [1–18].

Fractional derivative in Caputo sense:
Fractional derivative in Caputo sense is notified as follows [19–22]:

$$D^\alpha [f(t)] = I^{m-\alpha} D^m [f(t)] = \frac{1}{\Gamma(m-\alpha)} \int_0^t (t-\alpha)^{m-\alpha-1} f^m(x) dx$$

where $m - 1 < \alpha \leq m$.

Sumudu transform of Caputo derivative:
Sumudu transform of Caputo derivative is defined as follows [19]:

$$S[D_t^\alpha u(x,t)] = u^{-\alpha} S[u(x,t)] - \sum_{r=0}^{m-1} u^{-\alpha+r} u^r(0)$$

7.2 Main Text Implementation of the Scheme

Considered 2D nonlinear time-fractional coupled Burgers' equation is notified as follows:

$$D_t^\alpha u + uu_x + vu_y = \frac{1}{R}[u_{xx} + u_{yy}] \tag{7.1}$$

$$D_t^\alpha v + uv_x + vv_y = \frac{1}{R}[v_{xx} + v_{yy}] \tag{7.2}$$

Applying Sumudu transform upon eqn (7.1),

$$S[D_t^\alpha u(x,y,t)] = S\left[\frac{1}{R}[u_{xx} + u_{yy}] - [uu_x + vu_y]\right]$$

$$\Rightarrow u^{-\alpha} S[u(x,y,t)] - \sum_{r=0}^{m-1} u^{r-\alpha} u^r(0)$$

$$= S\left[\frac{1}{R}[u_{xx} + u_{yy}] - [uu_x + vu_y]\right]$$

$$\Rightarrow u^{-\alpha} S[u(x,y,t)]$$

$$= \sum_{r=0}^{m-1} u^{r-\alpha} u^r(0) + S\left[\frac{1}{R}[u_{xx} + u_{yy}] - [uu_x + vu_y]\right]$$

$$\Rightarrow S[u(x,y,t)]$$

$$= u^\alpha \sum_{r=0}^{m-1} u^{r-\alpha} u^r(0)$$

$$+ u^\alpha S\left[\frac{1}{R}[u_{xx} + u_{yy}] - [uu_x + vu_y]\right]$$

$$\Rightarrow u(x,y,t) = S^{-1}\left[u^\alpha \sum_{r=0}^{m-1} u^{r-\alpha} u^r(0)\right] + S^{-1}\left[u^\alpha S\left[\frac{1}{R}[u_{xx} + u_{yy}] - [uu_x + vu_y]\right]\right]. \tag{7.3}$$

By ADM,

$$u(x,y,t) = \sum_{n=0}^{\infty} u_n(x,y,t),\, uu_x = \sum_{n=0}^{\infty} A_n,\, vu_y = \sum_{n=0}^{\infty} B_n. \tag{7.4}$$

Using eqn (7.4) in eqn (7.3),

$$u_0(x,y,t) = S^{-1}\left[u^\alpha \sum_{r=0}^{m-1} u^{r-\alpha} u^r(0)\right] \tag{7.5}$$

$$u_{n+1}(x,y,t) = S^{-1}\left[u^\alpha S\left[\frac{1}{R}\left[\left(\sum_{n=0}^{\infty} u_n\right)_{xx} + \left(\sum_{n=0}^{\infty} u_n\right)_{yy}\right]\right.\right.$$
$$\left.\left. - \left[\sum_{n=0}^{\infty} A_n + \sum_{n=0}^{\infty} B_n\right]\right]\right], \tag{7.6}$$

where $n = 0, 1, 2, 3, \ldots$

$$u_1(x,y,t) = S^{-1}\left[u^\alpha S\left[\frac{1}{R}\left[(u_0)_{xx} + (u_0)_{yy}\right] - [A_0 + B_0]\right]\right] \tag{7.7}$$

$$u_2(x,y,t) = S^{-1}\left[u^\alpha S\left[\frac{1}{R}\left[(u_1)_{xx} + (u_1)_{yy}\right] - [A_1 + B_1]\right]\right] \tag{7.8}$$

$$u_3(x,y,t) = S^{-1}\left[u^\alpha S\left[\frac{1}{R}\left[(u_2)_{xx} + (u_2)_{yy}\right] - [A_2 + B_2]\right]\right]. \tag{7.9}$$

Applying Sumudu transform upon eqn (7.2)

$$S\left[D_t^\alpha v(x,y,t)\right] = S\left[\frac{1}{R}\left[v_{xx} + v_{yy}\right] - \left[uv_x + vv_y\right]\right]$$

$$\Rightarrow v^{-\alpha} S[v(x,y,t)] - \sum_{r=0}^{m-1} v^{r-\alpha} v^r(0) = S\left[\frac{1}{R}\left[v_{xx} + v_{yy}\right] - \left[uv_x + vv_y\right]\right]$$

$$\Rightarrow v^{-\alpha} S[v(x,y,t)] = \sum_{r=0}^{m-1} v^{r-\alpha} v^r(0) + S\left[\frac{1}{R}\left[v_{xx} + v_{yy}\right] - \left[uv_x + vv_y\right]\right]$$

$$\Rightarrow S[v(x,y,t)]$$
$$= v^\alpha \sum_{r=0}^{m-1} v^{r-\alpha} v^r(0)$$

$$+ v^\alpha S\left[\frac{1}{R}[v_{xx}+v_{yy}] - [uv_x + vv_y]\right]$$

$$\Rightarrow v(x,y,t) = S^{-1}\left[v^\alpha \sum_{r=0}^{m-1} v^{r-\alpha} v^r(0)\right] + S^{-1}\left[v^\alpha S\left[\frac{1}{R}[v_{xx}+v_{yy}] - [uv_x + vv_y]\right]\right]$$

(7.10)

By ADM,

$$v(x,y,t) = \sum_{n=0}^{\infty} v_n(x,y,t),\ uv_x = \sum_{n=0}^{\infty} C_n,\ vv_y = \sum_{n=0}^{\infty} D_n.$$

(7.11)

Using eqn (7.11) in eqn (7.10),

$$v_0(x,y,t) = S^{-1}\left[v^\alpha \sum_{r=0}^{m-1} v^{r-\alpha} v^r(0)\right]$$ (7.12)

$$v_{n+1}(x,y,t) = S^{-1}\left[v^\alpha S\left[\frac{1}{R}\left[\left(\sum_{n=0}^{\infty} v_n\right)_{xx} + \left(\sum_{n=0}^{\infty} v_n\right)_{yy}\right] - \left[\sum_{n=0}^{\infty} C_n + \sum_{n=0}^{\infty} D_n\right]\right]\right]$$

(7.13)

where $n = 0, 1, 2, 3, \ldots$

$$v_1(x,y,t) = S^{-1}\left[v^\alpha S\left[\frac{1}{R}\left[(v_0)_{xx} + (v_0)_{yy}\right] - [C_0 + D_0]\right]\right] \quad (7.14)$$

$$v_2(x,y,t) = S^{-1}\left[v^\alpha S\left[\frac{1}{R}\left[(v_1)_{xx} + (v_1)_{yy}\right] - [C_1 + D_1]\right]\right] \quad (7.15)$$

$$v_3(x,y,t) = S^{-1}\left[v^\alpha S\left[\frac{1}{R}\left[(v_2)_{xx} + (v_2)_{yy}\right] - [C_2 + D_2]\right]\right] \quad (7.16)$$

7.3 Examples and Calculation

Numerical Example 1:
Considered 2D time-fractional coupled Burgers' equation is notified as follows [23]:

7.3 Examples and Calculation

$$D_t^\alpha u + uu_x + vu_y = [u_{xx} + u_{yy}] \quad (7.17)$$
$$D_t^\alpha v + uv_x + vv_y = [v_{xx} + v_{yy}]. \quad (7.18)$$

I.C.:
$u(x,y,0) = x+y$ and $v(x,y,0) = x-y$

$$A_0 + B_0 = 2x, \ C_0 + D_0 = 2y$$
$$A_1 + B_1 = -4(x+y)\frac{t^{aa}}{2a}, \ C_1 + D_1 = -4(x-y)\frac{t^a}{\angle a}$$
$$A_2 + B_2 = 16x\frac{t^{2\alpha}}{\angle 2\alpha} + 4x\frac{t^{2\alpha}}{(\angle \alpha)^2}, \ C_2 + D_2 = 16y\frac{t^{2\alpha}}{\angle 2\alpha} + 4y\frac{t^{2\alpha}}{(\angle \alpha)^2}$$

From eqn (7.5): $u_0(x,y,t) = S^{-1}\left[u^\alpha \sum_{r=0}^{m-1} u^{r-\alpha} u^r(0)\right]$

$$\Rightarrow u_0(x,y,t) = S^{-1}\left[u^\alpha u^{-\alpha} u(0)\right] \Rightarrow u_0 = \boldsymbol{u}(0) = \boldsymbol{x+y}.$$

From eqn (7.12): $v_0(x,y,t) = S^{-1}\left[v^\alpha \sum_{r=0}^{m-1} v^{r-\alpha} v^r(0)\right]$

$$\Rightarrow v_0(x,y,t) = S^{-1}\left[v^\alpha v^{-\alpha} v(0)\right] \Rightarrow v_0 = v(0) = \boldsymbol{x-y}.$$

From eqn (7.7):
$$u_1(x,y,t) = S^{-1}\left[u^\alpha S\left[\frac{1}{R}\left[(u_0)_{xx} + (u_0)_{yy}\right] - [A_0 + B_0]\right]\right]$$
$$\Rightarrow u_1(x,y,t) = S^{-1}\left[u^\alpha S[-[2x]]\right] \Rightarrow u_1 = -[2x]S^{-1}\left[u^\alpha S[1]\right]$$
$$\Rightarrow u_1(x,y,t) = -[2x]S^{-1}\left[u^\alpha\right] \Rightarrow u_1(x,y,t) = -2x\frac{t^\alpha}{\angle\alpha}$$

From eqn (7.14); $v_1(x,y,t) = S^{-1}\left[v^\alpha S[1]\right]$

$$\Rightarrow v_1(x,y,t) = -2yS^{-1}\left[v^\alpha\right] \Rightarrow v_1(\boldsymbol{x,y,t}) = -2y\frac{t^\alpha}{\angle\alpha}$$

From eqn (7.8): $u_2(x,y,t) = S^{-1}\left[u^\alpha S\left[-\left[-4(x+y)\frac{t^\alpha}{\angle\alpha}\right]\right]\right]$

$$\Rightarrow u_2(x,y,t) = 4(x+y)S^{-1}\left[u^\alpha S\left[\frac{t^\alpha}{\angle\alpha}\right]\right] \Rightarrow u_2(x,y,t)$$
$$= 4(x+y)S^{-1}\left[u^\alpha u^\alpha\right]$$

$$\Rightarrow u_2(x,y,t) = 4(x+y)S^{-1}\left[u^{2\alpha}\right] \Rightarrow u_2(x,y,t) = 4(x+y)\frac{t^{2\alpha}}{\angle 2\alpha}.$$

From eqn (7.15): $v_2(x,y,t) = S^{-1}\left[v^\alpha S\left[-\left[-4(x-y)\frac{t^\alpha}{\angle\alpha}\right]\right]\right]$

$$\Rightarrow v_2(x,y,t) = 4(x-y)S^{-1}\left[v^\alpha S\left[\frac{t^\alpha}{\angle\alpha}\right]\right] \Rightarrow v_2(x,y,t)$$

$$= 4(x-y)S^{-1}\left[v^\alpha v^\alpha\right]$$

$$\Rightarrow v_2(x,y,t) = 4(x-y)S^{-1}\left[v^{2\alpha}\right] \Rightarrow v_2(\boldsymbol{x},\boldsymbol{y},\boldsymbol{t}) = 4(\boldsymbol{x}-\boldsymbol{y})\frac{t^{2\alpha}}{\angle 2\alpha}$$

From eqn (7.9): $u_3(x,y,t) = S^{-1}\left[u^\alpha S\left[-\left[16x\frac{t^{2\alpha}}{\angle 2\alpha} + 4x\frac{t^{2\alpha}}{(\angle\alpha)^2}\right]\right]\right]$

$$\Rightarrow u_3(x,y,t) = -S^{-1}\left[u^\alpha\left\{16xS\left(\frac{t^{2\alpha}}{\angle 2\alpha}\right) + 4xS\left(\frac{t^{2\alpha}}{(\angle\alpha)^2}\right)\right\}\right]$$

$$\Rightarrow u_3(x,y,t) = -S^{-1}\left[16xu^{3\alpha} + 4x\frac{\angle 2\alpha}{(\angle\alpha)^2}u^{3\alpha}\right]$$

$$\Rightarrow u_3(x,y,t) = -\left[16xS^{-1}\left[u^{3\alpha}\right] + 4x\frac{\angle 2\alpha}{(\angle\alpha)^2}S^{-1}\left[u^{3\alpha}\right]\right]$$

$$\Rightarrow u_3(x,y,t) = -\left[16x\frac{t^{3\alpha}}{\angle 3\alpha} + 4x\frac{\angle 2\alpha}{(\angle\alpha)^2}\frac{t^{3\alpha}}{\angle 3\alpha}\right] \Rightarrow u_3(x,y,t)$$

$$= -\left[16x + 4x\frac{\angle 2\alpha}{(\angle\alpha)^2}\right]\frac{t^{3\alpha}}{\angle 3\alpha}$$

From eqn (7.16): $v_3(x,y,t) = S^{-1}\left[u^\alpha S\left[-\left[16y\frac{t^{2\alpha}}{\angle 2\alpha} + 4y\frac{t^{2\alpha}}{(\angle\alpha)^2}\right]\right]\right]$

$$\Rightarrow v_3(x,y,t) = -S^{-1}\left[u^\alpha\left\{16yS\left(\frac{t^{2\alpha}}{\angle 2\alpha}\right) + 4yS\left(\frac{t^{2\alpha}}{(\angle\alpha)^2}\right)\right\}\right]$$

$$\Rightarrow v_3(x,y,t) = -S^{-1}\left[16yu^{3\alpha} + 4y\frac{\angle 2\alpha}{(\angle\alpha)^2}u^{3\alpha}\right]$$

$$\Rightarrow v_3(x,y,t) = -\left[16yS^{-1}\left[u^{3\alpha}\right] + 4y\frac{\angle 2\alpha}{(\angle\alpha)^2}s^{-1}\left[u^{3\alpha}\right]\right]$$

$$\Rightarrow v_3(x,y,t) = -\left[16y\frac{t^{3\alpha}}{\angle 3\alpha} + 4y\frac{\angle 2\alpha}{(\angle\alpha)^2}\frac{t^{3\alpha}}{\angle 3\alpha}\right] \Rightarrow v_3(x,y,t)$$

$$= -\left[16y + 4y\frac{\angle 2\alpha}{(\angle\alpha)^2}\right]\frac{t^{3\alpha}}{\angle 3\alpha}$$

$$u(x,y,t) = u_0(x,y,t) + u_1(x,y,t) + u_2(x,y,t) + u_3(x,y,t) + \cdots$$
$$\Rightarrow u(x,y,t) = (x+y) - 2x\frac{t^\alpha}{\angle\alpha} + 4(x+y)\frac{t^{2\alpha}}{\angle 2\alpha}$$
$$- \left[16x + 4x\frac{\angle 2\alpha}{(\angle\alpha)^2}\right]\frac{t^{3\alpha}}{\angle 3\alpha}$$

Considering $\alpha = 1$,

$$\Rightarrow u(x,y,t) = (x+y) - 2xt + 2(x+y)t^2 - [16x + 8x]\frac{t^3}{\angle 3}$$
$$\Rightarrow u(x,y,t) = x\left[1 + 2t^2 + 4t^4 + \cdots\right] + y\left[1 + 2t^2 + 4t^4 + \cdots\right]$$
$$- 2xt\left[1 + 2t^2 + 4t^4 + \cdots\right]$$
$$\Rightarrow u(x,y,t) = (x + y - 2xt)\left[1 + 2t^2 + 4t^4 + \cdots\right]$$
$$\Rightarrow u(x,y,t) = (x + y - 2xt)\left[1 - 2t^2\right]^{-1} \Rightarrow u = \frac{x + y - 2xt}{(1 - 2t^2)}$$

$$v(x,y,t) = v_0(x,y,t) + v_1(x,y,t) + v_2(x,y,t) + v_3(x,y,t) + \cdots$$
$$\Rightarrow v(x,y,t) = (x-y) - 2y\frac{t^\alpha}{\angle\alpha} + 4(x-y)\frac{t^{2\alpha}}{\angle 2\alpha}$$
$$- \left[16y + 4y\frac{\angle 2\alpha}{(\angle\alpha)^2}\right]\frac{t^{3\alpha}}{\angle 3\alpha}.$$

Considering $\alpha = 1$,

$$\Rightarrow v(x,y,t) = (x-y) - 2yt + 2(x-y)t^2 - [16y + 8y]\frac{t^3}{\angle 3}$$
$$\Rightarrow v(x,y,t) = x\left[1 + 2t^2 + 4t^4 + \cdots\right] - y\left[1 + 2t^2 + 4t^4 + \cdots\right]$$
$$- 2yt\left[1 + 2t^2 + 4t^4 + \cdots\right]$$
$$\Rightarrow v(x,y,t) = (x - y - 2yt)\left[1 + 2t^2 + 4t^4 + \cdots\right]$$
$$\Rightarrow v(x,y,t) = (x - y - 2yt)\left[1 - 2t^2\right]^{-1} \Rightarrow v(x,y,t) = \frac{x - y - 2yt}{(1 - 2t^2)}.$$

Exact solutions are exactly matched with [23].

Numerical Example 2:

2D fractional coupled Burgers' equation is notified as follows [23]:

$$D_t^\alpha u + \frac{uu_x}{x} + \frac{vu_y}{y} - \frac{(xu_x)_x}{x} - \frac{(yu_y)_y}{y} = (x^2 - y^2)e^t \qquad (7.19)$$

$$D_t^\alpha v + \frac{uv_x}{x} + \frac{vv_y}{y} - \frac{(xv_x)_x}{x} - \frac{(yv_y)_y}{y} = (x^2 - y^2) e^t \qquad (7.20)$$

I.C.: $u(x, y, 0) = x^2 - y^2$ and $v(x, y, 0) = x^2 - y^2$.

Applying Sumudu transform in eqn (7.19):

$S[D_t^\alpha u(x, y, t)]$

$$= S\left[(x^2 - y^2) e^t + \left[\frac{(xu_x)_x}{x} + \frac{(yu_y)_y}{y}\right] - \left[\frac{uu_x}{x} + \frac{vu_y}{y}\right]\right]$$

$S[D_t^\alpha u(x, y, t)]$

$$= S\left[(x^2 - y^2) e^t + \left[\frac{(u_x + xu_{xx})}{x} + \frac{(u_y + yu_{yy})}{y}\right]\right.$$

$$\left. - \left[\frac{uu_x}{x} + \frac{vu_y}{y}\right]\right]$$

$S[D_t^\alpha u(x, y, t)]$

$$= S\left[(x^2 - y^2) e^t + \left[\frac{u_x}{x} + \frac{u_y}{y}\right] + [u_{xx} + u_{yy}]\right.$$

$$\left. - \left[\frac{uu_x}{x} + \frac{vu_y}{y}\right]\right]$$

$$\Rightarrow u^{-\alpha} S[u(x, y, t)] - \sum_{r=0}^{m-1} u^{r-\alpha} u^r(0)$$

$$= S\left[(x^2 - y^2) e^t + \left[\frac{u_x}{x} + \frac{u_y}{y}\right] + [u_{xx} + u_{yy}] - \left[\frac{uu_x}{x} + \frac{vu_y}{y}\right]\right]$$

$$\Rightarrow S[u(x, y, t)] = u^\alpha \sum_{r=0}^{m-1} u^{r-\alpha} u^r(0)$$

$$+ u^\alpha \left[S\left[(x^2 - y^2) e^t\right] + S\left[\frac{u_x}{x} + \frac{u_y}{y}\right] + S[u_{xx} + u_{yy}]\right.$$

$$\left. - S\left[\frac{uu_x}{x} + \frac{vu_y}{y}\right]\right]$$

$$\Rightarrow u(x, y, t) = S^{-1}\left[u^\alpha \sum_{r=0}^{m-1} u^{r-\alpha} u^r(0)\right]$$

$$+ S^{-1}\left[u^\alpha \left[S\left[(x^2 - y^2) e^t\right] + S\left[\frac{u_x}{x} + \frac{u_y}{y}\right] + S[u_{xx} + u_{yy}]\right.\right.$$

$$\left.\left. - S\left[\frac{uu_x}{x} + \frac{vu_y}{y}\right]\right]\right]$$

7.3 Examples and Calculation

Using ADM, $uu_x = \sum_{n=0}^{\infty} A_n$ and $vu_y = \sum_{n=0}^{\infty} B_n$

$$u_0(x,y,t) = S^{-1}\left[u^a \sum_{r=0}^{m-1} u^{r-\alpha} u^r(0) + \frac{x^2 - y^2}{(1-u)} u^\alpha\right]$$

$$\Rightarrow u_0(x,y,t) = u(0) + (x^2 - y^2) S^{-1}\left[\frac{u}{1-u}\right] \quad \{\text{Considering } \alpha = 1\}$$

$$\Rightarrow u_0(x,y,t) = u(0) + (x^2 - y^2) S^{-1}\left[\frac{1-(1-u)}{1-u}\right]$$

$$\Rightarrow u_0(x,y,t) = u(0) + (x^2 - y^2) S^{-1}\left[\frac{1}{(1-u)} - 1\right]$$

$$\Rightarrow u_0(x,y,t) = (x^2 - y^2) + (x^2 - y^2)(e^t - 1) = (x^2 - y^2) e^t$$

$$\Rightarrow u_1(x,y,t) = S^{-1}\left[u^\alpha \left\{ S\left(\frac{(u_0)_x}{x} + \frac{(u_0)_y}{y}\right) + S\left((u_0)_{xx} + (u_0)_{yy}\right)\right.\right.$$
$$\left.\left. - S\left(\frac{A_0}{x} + \frac{B_0}{x}\right)\right\}\right]$$

$$\Rightarrow u_1(x,y,t) = -S^{-1}\left[u^\alpha S\left\{\left(\frac{A_0}{x} + \frac{B_0}{x}\right)\right\}\right] \Rightarrow u_1(x,y,t) = 0$$

Similarly, applying Sumudu transform in eqn (7.20):

$$S\left[D_t^\alpha v(x,y,t)\right]$$
$$= S\left[(x^2 - y^2) e^t + \left[\frac{(xv_x)_x}{x} + \frac{(yv_y)_y}{y}\right] - \left[\frac{uv_x}{x} + \frac{vv_y}{y}\right]\right]$$

$$\Rightarrow S\left[D_t^\alpha v(x,y,t)\right]$$
$$= S\left[(x^2 - y^2) e^t + \left[\frac{(v_x + xv_{xx})}{x} + \frac{(v_y + yv_{yy})}{y}\right]\right.$$
$$\left. - \left[\frac{uv_x}{x} + \frac{vv_y}{y}\right]\right]$$

$$\Rightarrow S\left[D_t^\alpha v(x,y,t)\right]$$
$$= S\left[(x^2 - y^2) e^t + \left[\frac{v_x}{x} + \frac{v_y}{y}\right] + [v_{xx} + v_{yy}]\right.$$

$$-\left[\frac{uv_x}{x} + \frac{vv_y}{y}\right]\Bigg]$$

$$\Rightarrow v^{-\alpha} S[v(x,y,t)] - \sum_{r=0}^{m-1} v^{r-\alpha} v^r(0)$$

$$= S\left[\left(x^2 - y^2\right)e^t + \left[\frac{v_x}{x} + \frac{v_y}{y}\right] + [v_{xx} + v_{yy}] - \left[\frac{uv_x}{x} + \frac{vv_y}{y}\right]\right]$$

$$\Rightarrow S[v(x,y,t)] = v^{\alpha} \sum_{r=0}^{m-1} v^{r-\alpha} v^r(0)$$

$$+ v^{\alpha}\left[S\left[\left(x^2 - y^2\right)e^t\right] + S\left[\frac{v_x}{x} + \frac{v_y}{y}\right] + S[v_{xx} + v_{yy}]\right.$$

$$\left. - S\left[\frac{uv_x}{x} + \frac{vv_y}{y}\right]\right]$$

$$\Rightarrow v(x,y,t) = S^{-1}\left[v^{\alpha} \sum_{r=0}^{m-1} v^{r-\alpha} v^r(0)\right]$$

Using ADM, $uv_x = \sum_{n=0}^{\infty} C_n$ and $vv_y = \sum_{n=0}^{\infty} D_n$

$$v_0(x,y,t) = S^{-1}\left[v^{\alpha} \sum_{r=0}^{m-1} v^{r-\alpha} v^r(0) + \frac{x^2 - y^2}{(1-u)} v^{\alpha}\right]$$

$$\Rightarrow v_0(x,y,t) = v(0) + \left(x^2 - y^2\right) S^{-1}\left[\frac{v}{1-v}\right] \quad \{\text{Considering } \alpha = 1\}$$

$$\Rightarrow v_0(x,y,t) = v(0) + \left(x^2 - y^2\right) S^{-1}\left[\frac{1-(1-v)}{1-v}\right]$$

$$\Rightarrow v_0(x,y,t) = v(0) + \left(x^2 - y^2\right) S^{-1}\left[\frac{1}{(1-v)} - 1\right]$$

$$\Rightarrow u_0(x,y,t) = \left(x^2 - y^2\right) + \left(x^2 - y^2\right)\left(e^t - 1\right) = \left(x^2 - y^2\right)e^t$$

$$\Rightarrow v_1(x,y,t) = S^{-1}\left[v^{\alpha}\left\{S\left(\frac{(v_0)_x}{x} + \frac{(v_0)_y}{y}\right) + S\left((v_0)_{xx} + (v_0)_{yy}\right)\right.\right.$$

$$\left.\left. - S\left(\frac{C_0}{x} + \frac{D_0}{x}\right)\right\}\right]$$

$$\Rightarrow v_1(x,y,t) = -S^{-1}\left[v^{\alpha} S\left\{\left(\frac{C_0}{x} + \frac{D_0}{x}\right)\right\}\right] \Rightarrow v_1(x,y,t) = 0$$

$$\Rightarrow u(x,y,t) = (x^2 - y^2)e^t \text{ and } v(x,y,t) = (x^2 - y^2)e^t$$

Exact solutions are exactly matched with [23].

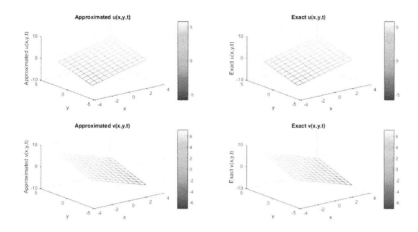

Figure 7.1 Compatibility of u and solution profiles at $t = 0.1$ for $N = 11$ for Example 1.

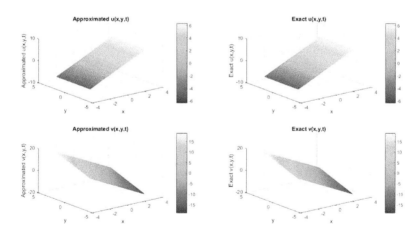

Figure 7.2 Compatibility of u and v solution profiles at $t = 0.5$ for $N = 51$ for Example 1.

7.4 Graphs and Discussion

Regarding Example 1, compatibility of approximate and exact solutions of u and v components at $t = 0.1$ for $N = 11$ is shown in Figure 7.1. Regarding Example 1, compatibility of approximate and exact solutions for u and v components at $t = 0.5$ for $N = 51$ is shown in Figure 7.2.

7.5 Concluding Remarks

The 2D nonlinear time-fractional coupled Burgers' equation has successfully been solved in the chapter, yielding approximate−exact solutions. The results are in the form of an infinite power series, which makes it simple to deduce the closed form of the solution. The outcomes of the present technique have produced results that are acceptable and accurate in light of the results reported in the literature. One of the most straightforward methods for solving fractional PDEs is the proposed regime. Additionally, using this method, approximate answers can be found for various fractional PDEs of complex characteristic.

Acknowledgements

I would like to thank Dr. Geeta Arora for giving an opportunity in the publication of this chapter.

References

[1] W.H. Su, D. Baleanu, X.J. Yang, H. Jafari, 'Damped wave equation and dissipative wave equation in fractal strings within the local fractional variational iteration method', Fixed Point Theory and Applications 2013, 1–11, 2013.

[2] H. Jafari, H.K. Jassim, 'Local fractional variational iteration method for nonlinear partial differential equations within local fractional operators', Applications and Applied Mathematics 10, 1055–1065, 2015.

[3] X.J. Yang, 'Local fractional functional analysis and its applications', Asian Academic, Hong Kong, China, 2011.

[4] S. Xu, X. Ling, Y. Zhao, H.K. Jassim, 'A novel schedule for solving the two-dimensional diffusion in fractal heat transfer', Thermal Science 19, S99–S103, 2015.

[5] H.K. Jassim, W.A. Shahab, 'Fractional variational iteration method to solve one dimensional second order hyperbolic telegraph equations', Journal of Physics: Conference Series 1032 (1) 1–9, 2018.

[6] X.J. Yang, J.A. Machad, H.M. Srivastava, 'A new numerical technique for solving the local fractional diffusion equation: Two-dimensional extended differential transform approach', Applied Mathematics and Computation 274 143– 151, 2016.

[7] H. Jafari, H.K. Jassim, F. Tchier, D. Baleanu, 'On the Approximate Solutions of Local Fractional Differential Equations with Local Fractional Operator', Entropy 18 1–12, 2016.

[8] H.K. Jassim, J. Vahidi, V.M. Ariyan, 'Solving Laplace Equation within Local Fractional Operators by Using Local Fractional Differential Transform and Laplace Variational Iteration Methods', Nonlinear Dynamics and Systems Theory 20 (4) 388–396, 2020.

[9] A.M. Yang, X.J. Yang, Z.B. Li, 'Local fractional series expansion method for solving wave and diffusion equations Cantor sets', Abstract and Applied Analysis 2013, 1– 5, 2013.

[10] H.K. Jassim, D. Baleanu, 'A novel approach for Korteweg-de Vries equation of fractional order', Journal of Applied Computational Mechanics 5 (2), 192–198, 2019.

[11] H.K. Jassim, S.A. Khafif, 'SVIM for solving Burger's and coupled Burger's equations of fractional order', Progress in Fractional Differentiation and Applications 7 (1), 1–6, 2021.

[12] H.A. Eaued, H.K. Jassim, 'A Novel Method for the Analytical Solution of Partial Differential Equations Arising in Mathematical Physics', IOP Conf. Series, Materials Science and Engineering 928 (042037), 1–16, 2020.

[13] C.G. Zhao et al, A.M. Yang, H. Jafari, H. Haghbin, 'The Yang-Laplace Transform for Solving the IVPs with Local Fractional Derivative', Abstract and Applied Analysis 2014, 1–5, 2014.

[14] Y. Zhang, X.J. Yang, C. Cattani, 'Local fractional homotopy perturbation method for solving nonhomogeneous heat conduction equations in fractal domain', Entropy 17, 6753–6764, 2015.

[15] H.K. Jassim, M.A. Shareef, 'On approximate solutions for fractional system of differential equations with Caputo-Fabrizio fractional operator', Journal of Mathematics and Computer science 23, 58–66, 2021.

[16] H.K. Jassim, H.A. Kadhim, 'Fractional Sumudu decomposition method for solving PDEs of fractional order', Journal of Applied and Computational Mechanics 7 (1), 302–311, 2021.

[17] D. Baleanu, H.K. Jassim, 'Exact Solution of Two-dimensional Fractional Partial Differential Equations', Fractal Fractional 4 (2), 21, https://doi.org/ 10.3390/fractalfract4020021, 2020.

[18] M.S. Hu, R.P. Agarwal, X.J. Yang, 'Local fractional Fourier series with application to wave equation in fractal vibrating', Abstract and Applied Analysis 2012, 1–7, 2012.

[19] A. Prakash, V. Verma, D. Kumar, J. Singh, 'Analytic study for fractional coupled Burger's equations via Sumudu transform method', *Nonlinear Engineering*, 7(4), 323-332, 2018.
[20] F. Mainardi, 'The fundamental solutions for the fractional diffusion-wave equation', *Applied Mathematics Letters*, 9(6), 23-28, 1996.
[21] Y. U. R. I. I. Luchko, R. Gorenflo, 'An operational method for solving fractional differential equations with the Caputo derivatives', *Acta Math. Vietnam*, 24(2), 207-233, 1999.
[22] F. Mainardi, 'On the initial value problem for the fractional diffusion-wave equation', *Waves and Stability in Continuous Media, World Scientific, Singapore*, 246-251, 1994.
[23] H. Eltayeb, I. Bachar, 'A note on singular two-dimensional fractional coupled Burgers' equation and triple Laplace Adomian decomposition method', *Boundary Value Problems*, *2020*(1), 1-17, 2020.

8

Physical and Dynamical Characterizations of the Wave's Propagation in Plasma Physics and Crystal Lattice Theory

Raghda A. M. Attia[1] and Mostafa M. A. Khater[2,3]

[1]Department of Basic Science,
Higher Technological Institute 10th of Ramadan City, Egypt
[2]School of Medical Informatics and Engineering,
Xuzhou Medical University, China
[3]Department of Basic Science,
Obour High Institute for Engineering and Technology, Egypt
Email: mostafa.khater2024@yahoo.com

Abstract

We provide exact soliton solutions in this study by addressing the Gilson–Pickering ($\mathcal{GP}$) issue using cutting-edge analytical and numerical methodologies. This concept describes the flow of waves through the medium in plasma physics and crystal lattice theory. The $\mathcal{GP}$ model may be used to create many evolution equations. The Fornberg–Whitham ($\mathcal{FW}$) equation, the Rosenau–Hyman ($\mathcal{RH}$) equation, and the Fuchssteiner–Fokas–Camassa–Holm ($\mathcal{FFCH}$) equation are a few examples. In this research effort, the Sardar sub-equation and He's variational iteration techniques are utilized to examine innovative wave characterizations in the disciplines of crystal lattice theory and plasma physics. A few replies may be described as formulae, and each one correlates to a graph. A conversation about how well the various tactics and alternative solutions fit together is a confirmation of the validity of the applications stated. Analytical solutions are investigated to determine whether they are stable using the Hamiltonian system as a test bed. The results show that the approach in issue is the most dependable option for solving nonlinear equations that emerge in mathematical physics.

Keywords: plasma physics, crystal lattice, computational simulations, soliton wave, stability.

AMS classification: 35Q60, 35E05, 35C08, 35Q51.

8.1 Introduction

To reduce it to its most fundamental components, plasma is an element. There is the existence of ions and/or electrons. Charged particles are a hallmark of plasma. It predominates in Sun-like stars and is consequently the most ubiquitous substance in the universe. It may go to faraway galaxies and possibly into clusters, producing plasma via the application of heat or an electromagnetic field. Electric currents in plasma are carried by charged particles, whose dynamics and macroscopic motion are regulated by electromagnetic fields and are particularly sensitive to external effects. Plasma's electromagnetic reactivity is exploited in plasma TVs and etching. Partially ionized plasma is created by neutral particles, which fluctuate with temperature and density. A partially ionized plasma, like neon or lightning, is responsible for these occurrences [1, 2].

Context is crucial during the plasma phase transition. The degree of ionization determines whether or not a material is a plasma. Crookes discovered a form of stuff known as plasma. Crookes addressed "radiant matter" in Sheffield in 1879, and Irving Langmuir researched plasma in the 1920s. In 1928, Langmuir invented the word "plasma" to characterize ionized gas. According to Lewi Tonks and Harold Mott-Smith, Langmuir purportedly compared it to blood plasma. How red and white blood cells and pathogens are conveyed by plasma inspired Langmuir [3, 4].

Because of their outstanding electrical conductivity, long-range electric and magnetic forces prevail in ionized materials. The particles in plasma, both positive and negative, are free to move around. Free particles may sense energies. The velocity of particles in plasma affects and is impacted by the fields of other charges. This governs the way individuals interact in groups. For all of matter, plasma is one of a kind. While both low-density plasma and "ionized gases" lack structure and volume, low-density plasma is not a gas. Following are the key distinctions [5, 6].

8.1 Introduction

As a consequence of the expansion nature of the electric force and the conductive character of plasma, the positive and negative charge densities are balanced. The plasma that is not neutral is either electrically charged or consists of only one species. Electricity may be present in plasma. Penning clouds of electrons and positrons in a trap is one example. Electrified particles may be discovered in plasma. Interactions between electrostatic particles of dust. Larger plasma granules. There is a lot of dirt in plasmas made in the lab. When plasma is ionized, it becomes charged. The plasma density is assessed with respect to the number of free electrons per space unit. In the universe, plasmas prevail over all other states of matter [7, 8].

Both the ionosphere and magnetosphere of Earth are packed with plasma. There is plasma from the surface of the sun to the heliopause. Low-density plasma may be found everywhere, including in the innards of all distant stars and over much of interplanetary and intergalactic space. In binary star systems, white dwarfs, neutron stars, and black holes all hold astrophysical plasmas. Astrophysical jets, such as M87s 5000-light-year jet, are coupled to plasma [9].

Synthetic plasma may be made by applying an electromagnetic field to a gas. Plasma may be formed in different ways, like its applications. All require energy to function. As the voltage rises, current strains the material, causing an electric spark. Insulator conducts. Electron-neutral gas collisions form the Townsend avalanche. Electrons that collide generate one ion and two electrons. Charged particles rise fast after 20 collisions [10].

Research, science, and industry utilize broad temperature and density ranges. Plasma spraying, microelectronic etching, metal cutting, and welding are applied in industrial and extractive metallurgy, automobile exhaust purification, fluorescent/luminescent lighting, and supersonic combustion engines. DC or low-frequency RF electric fields form non-thermal plasmas between metal electrodes. Fluorescent light bulbs create plasma [11].

13.56 MHz RF electric fields create plasmas akin to glow discharges. Glow discharges are weaker. Plasma etching and plasma-enhanced CVD are utilized in IC microfabrication. CCP and ICP are comparable RF-heated wave plasma. An example is helicon discharge: inflation high-power, high-temperature arc discharges. Several sources create it. Metallic use. Smelting Al2O3-containing materials yields aluminum. High-voltage electrode tips create non-thermal corona discharges. Ozone and particle precipitators employ it [12].

Non-thermal dielectric barrier discharge is formed by applying high voltages across microscopic gaps, where a non-conducting layer hinders plasma arcing. It is frequently mislabeled a 'Corona' shot, as are corona discharges. Plasma discharge minimizes vehicle drag. It is employed for web fabric. The release makes synthetic materials and plastics paint- and glue-friendly. The dielectric barrier discharge showed in the mid-1990s that low-temperature plasma kills bacteria. Plasma medicine was created from mammalian cell research. Dielectric barrier discharge was applied for low-temperature plasma jets. Plasma bullets emit plasma jets [13, 14].

Non-thermal plasma is created by supplying RF power to a powered and grounded electrode 1 cm apart. Helium or argon are employed to stabilize such discharges. High-side piezoelectric transformer non-thermal plasma. This generation is suitable for devices without a high-voltage power supply. In the 1960s, the world began studying magnetohydrodynamic converters to bring MHD power conversion to market with new commercial power plants. Supersonic and hypersonic aerodynamics were researched to passively manage flow around vehicles or projectiles, soften shock waves, limit heat transfer, and reduce drag [15].

Few "plasma technology" ionized gases are weakly ionized. "Cold" plasmas are weakly ionized gases. Magnetized non-thermal weakly ionized gases under magnetic fields require resistive magnetohydrodynamics with low magnetic Reynolds number, a challenging area of plasma physics needing dyadic tensors in seven–dimensional phase space. A key Hall parameter value creates electrothermal instability, impeding technological progress [16].

Plasma equations are fundamental; however, plasma behavior is varied and sophisticated; unexpected behavior from a simple model indicates a complex system. Such systems fall between structured and chaotic activity and cannot be described by mathematics or chance. Plasma complexity generates magnetic spatial properties. Icon parts are sharp, intermittent, or fractal. These lab-studied characteristics are now ubiquitous [17].

Birkeland currents occur in plasma balls, auroras, lightning, electric arcs, solar flares, and supernova remnants. High current density and magnetic field interaction may form a magnetic rope. Filamentation is a self-focusing laser. High powers demand a nonlinear refraction index. It causes a greater index of refraction in the laser's brightest center, providing feedback that focuses it more. Higher laser brightness causes plasma. A lower plasma refraction index defocuses the laser beam. Refractive index focusing and defocusing plasma produce a micrometers-to-kilometers-long filament. Filamentation-generated plasma has low ion density due to electron defocusing [18, 19].

Thermal impermeable plasma may be pushed like a solid. In the 1960s and 1970s, Hannes Alfvén's group studied cold gas and hot plasma to shelter fusion plasma from reactor walls. The external magnetic fields in this arrangement generated plasma kink instabilities and considerable wall heat loss. In 2013, scientists developed stable impermeable plasma without magnetic confinement. High pressure makes plasma spectroscopy hard; yet, its passive impact on nanostructure creation predicts successful captivity. Maintaining impermeability for a few tens of seconds led to a significant secondary heating mode, changing processes, and nanomaterial formation kinetics [20, 21].

In this context, we study the $\mathcal{GP}$ model that is given in the following formula [22, 23, 24]:

$$-r_1 \frac{\partial^3 \mathcal{U}}{\partial x \partial x \partial t} - r_4 \frac{\partial \mathcal{U}}{\partial x} \frac{\partial^2 \mathcal{U}}{\partial x \partial x} - \mathcal{U} \frac{\partial^3 \mathcal{U}}{\partial x \partial x \partial x} - r_3 \mathcal{U} \frac{\partial \mathcal{U}}{\partial x} + 2 r_2 \frac{\partial \mathcal{U}}{\partial x} + \frac{\partial \mathcal{U}}{\partial t} = 0, \tag{8.1}$$

where r_1, r_2, r_3, and r_4 are arbitrary parameters while $\mathcal{U}$ describes wave propagation in plasma physics and crystal lattice theory [25]. Using the next wave transformation $\mathcal{U}(x, t) = \mathcal{Q}(\gamma)$, $\gamma = ct + \lambda x$, where c, and λ are arbitrary constants to be determined through the method's steps, and integrating the obtained equation once with zero integration constant, leads to

$$\mathcal{Q}(c + 2\lambda r_2) - \frac{1}{2} \lambda \left(\lambda \left(2 \mathcal{Q}''(c r_1 + \lambda \mathcal{Q}) + \lambda (r_4 - 1) (\mathcal{Q}')^2 \right) \right. \\ \left. + r_3 \mathcal{Q}^2 \right) = 0. \tag{8.2}$$

Balancing eqn (8.2)'s terms along with Khater II method's auxiliary equation $\phi'(\xi)^2 \to \gamma_3 \phi(\xi)^4 + \gamma_2 \phi(\xi)^3 + \gamma_1 \phi(\xi)^2$ [26, 27], where γ_1, γ_2, and γ_3 [26, 27, 28] are arbitrary constants, leads to $N = 2$. Thus, the investigated model's general solution according to the employed analytical scheme is formulated as follows:

$$\mathcal{Q}(\gamma) = \sum_{i=1}^{n} \left(a_i \phi(\xi)^i + \frac{b_i}{\phi(\xi)^i} \right) + a_0$$

$$= a_2 \phi(\xi)^2 + a_1 \phi(\xi) + a_0 + \frac{b_2}{\phi(\xi)^2} + \frac{b_1}{\phi(\xi)}, \tag{8.3}$$

where a_0, a_1, a_2, b_1, and b_2 are arbitrary constants to be determined later.

The remainder of the chapter is organized as follows. Section 8.2 studies some novel analytical solutions of the investigated model and its accuracy by employing Sardar sub–equation method and He's variational iteration technique. Section 8.3 explains the paper's contributions and the results' novelty. Section 8.4 gives the conclusion of the whole study.

8.2 GP model's Traveling Wave Solutions

Here, we investigate the analytical and approximate solutions of the $\mathcal{GP}$ model through the Sardar sub–equation method and He's variational iteration technique.

8.2.1 Solitary wave solutions

Handling the investigated model by using the Sardar sub-equation method gets the following values of the above-mentioned parameters:

$$a_1 \to 0, a_2 \to 0, b_2 \to \frac{b_1\gamma_1}{\gamma_2}, \lambda \to \frac{\sqrt{r_3}}{2\sqrt{-(\gamma_1(r_4+1))}},$$

$$r_1 \to \frac{\sqrt{r_3}(r_4-1)(4a_0\gamma_1 - b_1\gamma_2)(a_0\gamma_2 - b_1\gamma_3)}{2c\sqrt{-(\gamma_1(r_4+1))}(8a_0\gamma_1\gamma_2 - b_1(\gamma_2^2 + 4\gamma_1\gamma_3))},$$

$$r_2 \to \frac{r_3\left(-8a_0 b_1 \gamma_1 \gamma_2 \left(\gamma_2^2 + 4\gamma_1\gamma_3\right)(r_4+1) + 32 a_0^2 \gamma_1^2 \gamma_2^2 (r_4+1) + b_1^2 \left(8\gamma_1\gamma_3\gamma_2^2 + \left(\gamma_2^4 + 16\gamma_1^2\gamma_3^2\right)r_4\right)\right)}{16\gamma_1\gamma_2(r_4+1)(8a_0\gamma_1\gamma_2 - b_1(\gamma_2^2 + 4\gamma_1\gamma_3))}$$

$$-\frac{c\sqrt{-(\gamma_1(r_4+1))}}{\sqrt{r_3}}.$$

In combination with the above-mentioned parameters, the analytical technique leads to the following soliton wave solutions.

For $\gamma_1 \neq 0$, $\gamma_2 \neq 0$, and $\gamma_3 \neq 0$, we get

$$\mathcal{U}(x,t)_I(x,t)$$

$$= a_0 + \frac{b_1}{8\gamma_1\gamma_2}\left(-(\gamma_2^2 - 4\gamma_1\gamma_3)\cosh\left(\sqrt{\gamma_1}\left(2ct + \frac{\sqrt{r_3}x}{\sqrt{-(\gamma_1(r_4+1))}}\right.\right.\right.$$

$$\left.\left.\left.-2\varsigma\right)\right) - \gamma_2^2 - 4\gamma_1\gamma_3\right). \tag{8.4}$$

For $\gamma_1 = \frac{\gamma_2^2}{4}$, $\gamma_2 \neq 0$, and $\gamma_3 \neq 0$, we get

$$\mathcal{U}(x,t)_{II}(x,t) \tag{8.5}$$

$$= \frac{a_1\left(\gamma_2\operatorname{sech}^2\left(\frac{\Omega}{2}\right) + 4\sqrt{\gamma_2^2(\gamma_3-1)\sinh^6\left(\frac{\Omega}{2}\right)\operatorname{csch}^4(\Omega)}\right)}{2\gamma_3\tanh^2\left(\frac{\Omega}{2}\right) - 2} + a_0$$

$$+ \frac{b_1\gamma_2\left(2\gamma_3\tanh^2\left(\frac{\Omega}{2}\right) - 2\right)^2}{4\left(\gamma_2\operatorname{sech}^2\left(\frac{\Omega}{2}\right) + 4\sqrt{\gamma_2^2(\gamma_3-1)\sinh^6\left(\frac{\Omega}{2}\right)\operatorname{csch}^4(\Omega)}\right)^2}$$

$$+ \frac{b_1\left(2\gamma_3\tanh^2\left(\frac{\Omega}{2}\right)-2\right)}{\gamma_2\mathrm{sech}^2\left(\frac{\Omega}{2}\right)+4\sqrt{\gamma_2^2(\gamma_3-1)\sinh^6\left(\frac{\Omega}{2}\right)\mathrm{csch}^4(\Omega)}}, \tag{8.6}$$

where $\Omega = \gamma_2\left(ct + \frac{\sqrt{r_3}x}{\sqrt{-(\gamma_2^2(r_4+1))}} + 2\varsigma\right)$.

For $\gamma_1 \neq 0$, $\gamma_2 \neq 0$, and $\gamma_3 = \gamma_1\gamma_2$, we get

$$\mathcal{U}(x,t)_{III}(x,t)$$

$$= \frac{2a_1\gamma_1\left(\gamma_1\sqrt{\frac{(4\gamma_1^2-\gamma_2)\gamma_2\tanh^2(\Omega_1)\mathrm{sech}^2(\Omega_1)}{\gamma_1^2}} - \gamma_2\mathrm{sech}^2(\Omega_1)\right)}{\gamma_2\left(\gamma_2 - 4\gamma_1^2\tanh^2(\Omega_1)\right)} + a_0$$

$$+ \frac{b_1\gamma_2\left(\gamma_2 - 4\gamma_1^2\tanh^2(\Omega_1)\right)^2}{4\gamma_1\left(\gamma_2\mathrm{sech}^2(\Omega_1) - \gamma_1\sqrt{\frac{(4\gamma_1^2-\gamma_2)\gamma_2\tanh^2(\Omega_1)\mathrm{sech}^2(\Omega_1)}{\gamma_1^2}}\right)^2} \tag{8.7}$$

$$+ \frac{b_1\gamma_2\left(\gamma_2 - 4\gamma_1^2\tanh^2(\Omega_1)\right)}{2\gamma_1\left(\gamma_1\sqrt{\frac{(4\gamma_1^2-\gamma_2)\gamma_2\tanh^2(\Omega_1)\mathrm{sech}^2(\Omega_1)}{\gamma_1^2}} - \gamma_2\mathrm{sech}^2(\Omega_1)\right)},$$

where $\Omega_1 = \sqrt{\gamma_1}\left(ct + \frac{\sqrt{r_3}x}{2\sqrt{-(\gamma_1(r_4+1))}} + \varsigma\right)$.

For $\gamma_2^2 - 4\gamma_1\gamma_3 = 0$, we get

$$\mathcal{U}(x,t)_{B,II}(x,t)$$

$$= \frac{a_1\gamma_2 e^{\gamma_2\varsigma}}{\exp\left(\frac{\gamma_2\left(ct+\frac{\sqrt{r_3}x}{2\sqrt{-(\gamma_1(r_4+1))}}\right)}{2\sqrt{\gamma_3}}\right) - 2\gamma_3 e^{\gamma_2\varsigma}} + a_0$$

$$+ \frac{b_1\gamma_1 e^{-2\gamma_2\varsigma}}{\gamma_2^3}\left(\exp\left(\frac{\gamma_2\left(ct+\frac{\sqrt{r_3}x}{2\sqrt{-(\gamma_1(r_4+1))}}\right)}{2\sqrt{\gamma_3}}\right) - 2\gamma_3 e^{\gamma_2\varsigma}\right)^2 \tag{8.8}$$

$$+ \frac{b_1 e^{\gamma_2(-\varsigma)}}{\gamma_2}\left(\exp\left(\frac{\gamma_2\left(ct+\frac{\sqrt{r_3}x}{2\sqrt{-(\gamma_1(r_4+1))}}\right)}{2\sqrt{\gamma_3}}\right) - 2\gamma_3 e^{\gamma_2\varsigma}\right).$$

8.2.2 Solution's accuracy

He's variational iteration method is demonstrated here with a general nonlinear differential equation [29, 30]:

$$\mathscr{L}\mathcal{U}(x,t) + \mathscr{N}\mathcal{U}(x,t) = g(x,t), \tag{8.9}$$

where the symbols $\mathscr{L}$, $\mathscr{N}$, and $g(x,t)$ denote linear operators, nonlinear operators, and known analytical functions, respectively. Thus, the correction functional can be given by

$$\mathcal{U}_{\Xi+1}(x,t) = \mathcal{U}_\Xi(x,t) + \int_0^t \beta \left(\mathscr{L}\mathcal{U}_\Xi(x,\psi) + \mathscr{N}\tilde{\mathcal{U}}_\Xi(x,\psi) - g(x,\psi) \right) d\psi. \tag{8.10}$$

In this case, β denotes a general Lagrange multiplier that can be identified optimally using variational theory, and $\tilde{\mathcal{U}}_\Xi$ represents a restricted variation, i.e., $\beta \tilde{\mathcal{U}}_\Xi = 0$. The stationary conditions

$$\begin{cases} 1 + \beta = 0, \\ \beta' = 0. \end{cases} \tag{8.11}$$

This in turn gives

$$\beta = -1. \tag{8.12}$$

The following values for the analytical, numerical, and absolute differences between these two values are found using He's variational iteration technique for investigating the $\mathcal{GP}$ model's approximate solution Table 8.1.

8.3 Soliton Solution's Novelty

This section details the results of our study and discusses how they add to the existing body of knowledge. The researchers have successfully used two analytical and numerical methodologies for the model under study and have collected a sizable number of computational and approximative solutions. Numerical diagrams (Figures 8.1–8.4) have been produced to illustrate these solutions, which display phenomena including periodic, singular, and type waves in two-, three-, density-, and polar-graphs, respectively. When our constructed solutions are compared to those that have been recently published, it

8.3 Soliton Solution's Novelty 125

Table 8.1 Analytical and approximate values for $x \in \{-30, -29, -28, \ldots, -2, -1, 0\}$.

Value of ξ	Analytical Sol.	Approximate Sol.	|Error|	Value of ξ	Analytical Sol.	Approximate Sol.	|Error|
−30	6.96934527470102	6.96934527272189	5.70874014726996E-10	1	6.99011654104448	6.99011654072241	3.2207037037324E-10
−29	6.38801847494115	6.38801847773432	2.7931728041608E-09	2	6.70277711905646	6.70277712094969	1.89323667854069E-09
−28	5.56487370229222	5.56487370536108	3.068885894474362E-09	3	5.88423207836363	5.88423208180641	3.44277673036686E-09
−27	5.25670646460106	5.25670646405813	5.42939027070588E-10	4	5.28704795446512	5.28704795571876	1.25363719405414E-09
−26	5.74684429901489	5.746844429562512	3.3897631368518E-09	5	5.46027307681753	5.46027307413182	2.68571476169654E-09
−25	6.58465674096006	6.58465673868477	2.2752866257747E-09	6	6.24464506679864	6.24464506371361	3.08502734469585E-09
−24	6.99986289508222	6.99986289504444	3.78159725755722E-11	7	6.91901593413657	6.91901593319499	9.4158105363015E-10
−23	6.61072413810274	6.6107241402988	2.19606022255903E-09	8	6.86337221342143	6.86337221466254	1.24110766108743E-09
−22	5.77501284858323	5.77501285199495	3.41171180195943E-09	9	6.12887248816455	6.12887248812872	3.26417115559252E-09
−21	5.26107813202786	5.261078132272389	6.96037005809558E-10	10	5.39081241158209	5.39081241387785	2.29576535559772E-09
−20	5.54142919670625	5.54142919370721	2.99904012734942E-09	11	5.32776102121151	5.32776101943873	1.77278547397464E-09
−19	6.35831256665827	6.35831256379861	2.85966095248114E-09	12	5.99768747128315	5.99768746788561	3.39754446798679E-09
−18	6.96068943880071	6.96068943815255	6.48158859917203E-10	13	6.78466447312525	6.78466447153866	1.58659130278238E-09
−17	6.79473729488908	6.7947372964348	1.5457226609783E-09	14	6.96514900118405	6.96514900118366	6.09468919776646E-10
−16	6.01303177070821	6.01303177049436	3.38615269157572E-09	15	6.37320441144126	6.37320441426798	2.8267113094671E-09
−15	5.33426932017027	5.33426932200855	1.83828152700016E-09	16	5.553061834111682	5.553061834111682	3.03468628004566E-09
−14	5.38250101002642	5.382501007788	2.2384210041082E-09	17	5.25875656559809	5.25875656497843	6.19658102607445E-10
−13	6.11338284705475	6.11338284377077	3.28397664617341E-09	18	5.76087515881689	5.760875151541542	3.40146844024503E-09
−12	6.85494543964421	6.85494543964421	1.28148187511738E-09	19	6.59776451837091	6.59776451613513	2.2357786733096E-09
−11	6.92539956272543	6.92539953626753	9.02096175536826E-10	20	7	7	0
−10	6.25997001778711	6.25997002084451	3.05739789041581E-09	21	6.59776451689834	6.59776451913411	2.23577795614802E-09
−9	5.47044966094145	5.47044966367353	2.73208922152435E-09	22	5.7608751722562	5.76087512062708	3.40146844024503E-09
−8	5.28271986721267	5.282719886603153	1.1811410690295E-09	23	5.25875656535113	5.25875656597079	6.1965899075865E-10
−7	5.86937854319472	5.86937853975153	3.44319062151044E-09	24	5.55306183240654	5.55306182937186	3.03468539186724E-09
−6	6.6910544077048	6.69105440577082	1.93398363999127E-09	25	6.37320441311938	6.37320441029267	2.8267113094671E-09
−5	6.99230246026473	6.99230246054877	2.84035017728002E-10	26	6.96514900106356	6.96514900045409	6.09468031598226E-10
−4	6.49615663062749	6.49615663315276	2.5252688828914E-09	27	6.78466447197553	6.78466447356212	1.5865921909608E-09
−3	5.65877110646783	5.65877110974523	3.27739879679712E-09	28	5.99768746955177	5.99768747294932	3.39754446798679E-09
−2	5.25003427689695	5.25003427693591	3.89670518075036E-11	29	5.3277610204903	5.327761022262309	1.772786215306E-09
−1	5.64573689903584	5.64573689478136	3.25447668814149E-09	30	5.39081241253412	5.39081241023836	2.29576535559772E-09

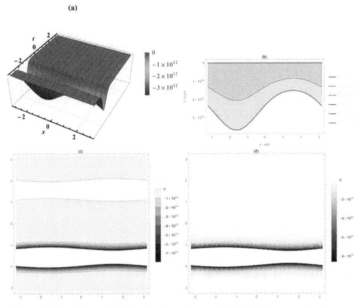

Figure 8.1 Numerical simulations of eqn (8.4) in distinct graphs' type.

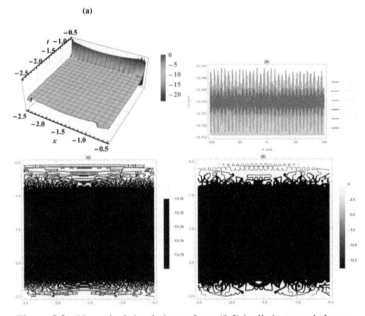

Figure 8.2 Numerical simulations of eqn (8.5) in distinct graphs' type.

8.3 Soliton Solution's Novelty 127

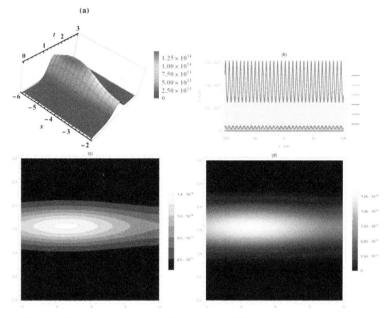

Figure 8.3 Numerical simulations of eqn (8.7) in distinct graphs' type.

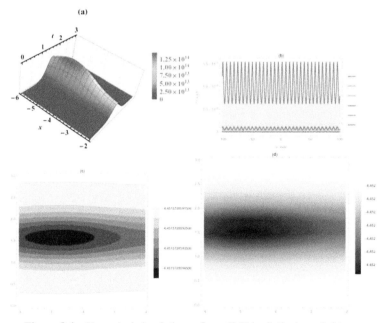

Figure 8.4 Numerical simulations of eqn (8.8) in distinct graphs' type.

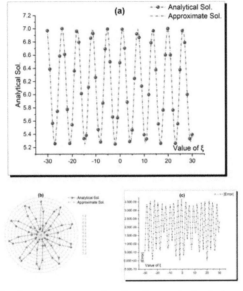

Figure 8.5 Matching between computational and numerical solutions based on Table 8.1.

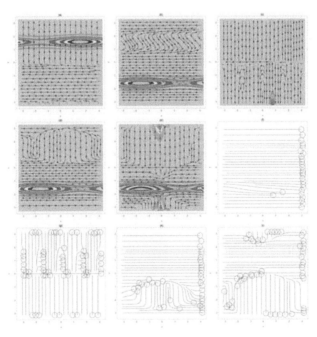

Figure 8.6 In a crystal, vibrations are caused by waves passing through its atoms.

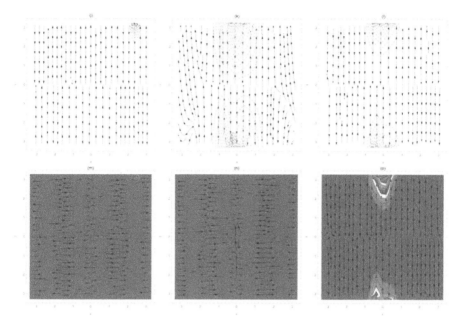

Figure 8.7 An atom in a crystal vibrates as waves pass through it.

is possible to demonstrate the novelty of some of our results. At the same time, Figure 8.5 displays an approximation of the answer and the correctness of that approximation based on He's variational iteration method's solution to the examined model. However, by choosing appropriate values for the parameters mentioned above, we get that findings are consistent with those of previously published works [23, 24, 25]. The waves that pass through a crystal and generate vibrations in its atoms may be determined using the solutions developed in the preceding paragraphs; also, a crystalline solid vibrates when a wave travels through its atoms in plasma. The outcomes are shown in Figures 8.6, and 8.7.

8.4 Conclusion

The authors of this chapter tried to solve the mathematical $\mathcal{GP}$ problem by coming up with a complete set of equations that led to many different solutions for soliton waves. For the theory of crystal lattices and to study how waves move in plasma, this kind of solution is needed. Each of these groups has been looked at in detail, and the connections between them have been shown using a variety of visuals. The importance of our research was determined by comparing our results to those of other published studies related to the same issue and the same problem. Mathematica 13.1 validated was used to validate all solutions by re-inserting them into the original model that generated them. The scenario called for various approaches to the problem to be considered and a solution to be discovered.

Declarations

Ethics approval and consent to participate

Not applicable.

Consent for publication

Not applicable.

Availability of data and material

The data that support the findings of this study are available from the corresponding author upon reasonable request.

Competing interests

The authors declare that they have no competing interests.

Funding

No fund has been received for this paper.

Authors' contributions

All authors contributed equally to the writing of this paper. All authors read and approved the final manuscript.

Acknowledgments

I greatly thank the journal staff (Editors and Reviewers) for their support and help.

References

[1] M. M. Khater, R. A. Attia, D. Lu, Modified auxiliary equation method versus three nonlinear fractional biological models in present explicit wave solutions, Mathematical and Computational Applications 24 (1) (2018) 1.

[2] M. M. Khater, M. S. Mohamed, R. A. Attia, On semi analytical and numerical simulations for a mathematical biological model; the time-fractional nonlinear kolmogorov–petrovskii–piskunov (kpp) equation, Chaos, Solitons & Fractals 144 (2021) 110676.

[3] M. Khater, C. Park, D. Lu, R. A. Attia, Analytical, semi-analytical, and numerical solutions for the cahn–allen equation, Advances in Difference Equations 2020 (1) (2020) 1–12.

[4] M. Khater, C. Park, D. Lu, R. A. Attia, Analytical, semi-analytical, and numerical solutions for the cahn–allen equation, Advances in Difference Equations 2020 (1) (2020) 1–12.

[5] M. M. Khater, D. Lu, R. A. Attia, Lump soliton wave solutions for the (2+ 1)-dimensional konopelchenko–dubrovsky equation and kdv equation, Modern Physics Letters B 33 (18) (2019) 1950199.

[6] C. Yue, M. Khater, R. A. Attia, D. Lu, The plethora of explicit solutions of the fractional ks equation through liquid–gas bubbles mix under the thermodynamic conditions via atangana–baleanu derivative operator, Advances in Difference Equations 2020 (1) (2020) 1–12.

[7] M. M. Khater, A. E.-S. Ahmed, M. El-Shorbagy, Abundant stable computational solutions of atangana–baleanu fractional nonlinear hiv-1 infection of cd4+ t-cells of immunodeficiency syndrome, Results in Physics 22 (2021) 103890.

[8] M. M. Khater, A. Mousa, M. El-Shorbagy, R. A. Attia, Analytical and semi-analytical solutions for phi-four equation through three recent schemes, Results in Physics 22 (2021) 103954.

[9] M. M. Khater, R. A. Attia, D. Lu, Explicit lump solitary wave of certain interesting (3+ 1)-dimensional waves in physics via some recent traveling wave methods, Entropy 21 (4) (2019) 397.

[10] M. M. Khater, Diverse solitary and jacobian solutions in a continually laminated fluid with respect to shear flows through the ostrovsky equation, Modern Physics Letters B 35 (13) (2021) 2150220.

[11] M. M. Khater, Abundant breather and semi-analytical investigation: On high-frequency waves' dynamics in the relaxation medium, Modern Physics Letters B 35 (22) (2021) 2150372.

[12] M. M. Khater, R. A. Attia, C. Park, D. Lu, On the numerical investigation of the interaction in plasma between (high & low) frequency of (langmuir & ion-acoustic) waves, Results in Physics 18 (2020) 103317.

[13] M. M. Khater, D. Lu, Analytical versus numerical solutions of the nonlinear fractional time–space telegraph equation, Modern Physics Letters B 35 (19) (2021) 2150324.

[14] M. M. Khater, Nonparaxial pulse propagation in a planar waveguide with kerr–like and quintic nonlinearities; computational simulations, Chaos, Solitons & Fractals 157 (2022) 111970.

[15] M. M. Khater, Lax representation and bi-hamiltonian structure of nonlinear qiao model, Modern Physics Letters B 36 (07) (2022) 2150614.

[16] M. M. Khater, Two-component plasma and electron trapping's influence on the potential of a solitary electrostatic wave with the dust-ion-acoustic speed, Journal of Ocean Engineering and Science (2022).

[17] M. M. Khater, D. Lu, Diverse soliton wave solutions of for the nonlinear potential kadomtsev–petviashvili and calogero–degasperis equations, Results in Physics 33 (2022) 105116.

[18] M. M. Khater, A. M. Alabdali, A. Mashat, S. A. Salama, et al., Optical soliton wave solutions of the fractional complex paraxial wave dynamical model along with kerr media, FRACTALS (fractals) 30 (05) (2022) 1–17.

[19] M. M. Khater, In solid physics equations, accurate and novel soliton wave structures for heating a single crystal of sodium fluoride, International Journal of Modern Physics B (2022) 2350068.

[20] M. M. Khater, De broglie waves and nuclear element interaction; abundant waves structures of the nonlinear fractional phi-four equation, Chaos, Solitons & Fractals 163 (2022) 112549.

[21] M. M. Khater, Novel computational simulation of the propagation of pulses in optical fibers regarding the dispersion effect, International Journal of Modern Physics B (2022) 2350083.

[22] H. K. Park, M. J. Choi, M. Kim, M. Kim, J. Lee, D. Lee, W. Lee, G. Yun, Advances in physics of the magneto-hydro-dynamic and turbulence-based instabilities in toroidal plasmas via 2-D/3-D visualization, Reviews of Modern Plasma Physics 6 (1) (2022) 18. doi: 10.1007/s41614-022-00076-2.

[23] P. Attri, K. Koga, T. Okumura, F. L. Chawarambwa, T. E. Putri, Y. Tsukada, K. Kamataki, N. Itagaki, M. Shiratani, Treatment of organic wastewater by a combination of non-thermal plasma and catalyst: a

review, Reviews of Modern Plasma Physics 6 (1) (2022) 17. doi: 10.1007/s41614-022-00077-1.

[24] S. Bhattacharjee, A. R. Baitha, A. Nanda, S. Hunjan, S. Bhattacharjee, Physics of plasmas confined by a dipole magnet: insights from compact experiments driven at steady state, Reviews of Modern Plasma Physics 6 (1) (2022) 16. doi:10.1007/s41614-022-00078-0.

[25] M. Zhou, Z. Zhong, X. Deng, Kinetic properties of collisionless magnetic reconnection in space plasma: in situ observations, Reviews of Modern Plasma Physics 6 (1) (2022) 15. doi:10.1007/s41614-022-00079-z.

[26] W. Zheng-Xiong, L. Tong, W. Lai, Nonlinear evolution and control of neo-classical tearing mode in reversed magnetic shear tokamak plasmas, Reviews of Modern Plasma Physics 6 (1) (2022) 14. doi:10.1007/s41614-022-00074-4.

[27] K. Takeda, K. Ishikawa, M. Hori, Wide range applications of process plasma diagnostics using vacuum ultraviolet absorption spectroscopy, Reviews of Modern Plasma Physics 6 (1) (2022) 13. doi:10.1007/s41614-022-00075-3.

[28] L. Wang, Interplanetary energetic electrons observed in Earth's polar cusp/cap/lobes, Reviews of Modern Plasma Physics 6 (1) (2022) 12. doi:10.1007/s41614-022-00073-5.

[29] W. Masood, H. A. Shah, M. N. S. Qureshi, Trapping in quantum plasmas: a review, Reviews of Modern Plasma Physics 6 (1) (2022) 11. doi:10.1007/s41614-022-00072-6.

[30] P. Rodriguez-Fernandez, C. Angioni, A. E. White, Local transport dynamics of cold pulses in tokamak plasmas, Reviews of Modern Plasma Physics 6 (1) (2022) 10. doi:10.1007/s41614-022-00071-7.

9

Numerical Solution of Fractional-order One-dimensional Differential Equations by using Laplace Transform with the Residual Power Series Method

Rajendra Pant[1], Geeta Arora[1], Manik Rakhra[2], and Masoumeh Khademi[3]

[1,*]Department of Mathematics, School of Chemical Engineering and Physical Sciences, Lovely Professional University, India
[2]School of Computer Science and Engineering, Lovely Professional University, India
[3]Department of Mathematics, Hamedan Branch, Islamic Azad University, Iran
E-mail: rpant2036@gmail.com; geetadma@gmail.com; rakhramanik786@gmail.com; dr.amonaft@gmail

Abstract

The Laplace transform with residual power series method is one of the efficient and reliable methods for the solution of fractional-order linear as well as nonlinear differential equations. The aim of this work is to find the solution of one-dimensional fractional-order differential equation by using Laplace transform with residual power series method. The comparison between exact solution and approximate solutions of these equations by Laplace residual power series method is determined. The unknown coefficients of the series are obtained by using Laplace transform with residual power series method. This method reduces the size of computational works and the solution is obtained in series form.

Keywords: Laplace transforms, Schrödinger differential equation, residual power series method, Laplace residual function.

9.1 Introduction

One of the generalized forms of the classical differential equations is the fractional differential equations that have the considerable applications in different branches of sciences from last decades and some fundamental facts and numerous definitions of fractional calculus are specified in a large number of books [1]–[4]. However, there are so many analytical methods for numerical solutions of fractional-order differential equations, and out of them, most common and applicable methods are presented in the literature [5]–[15]. The Laplace transform with residual power series method is used for solution of fractional-order differential equations manually.

The residual power series method (RPSM) is one of the efficient and reliable approaches for the solution of linear as well as nonlinear differential equations of fractional order in closed form of solution of well-known functions [16]. In case of linear fractional differential equations, sometimes, it is impossible to find such solutions and it is very difficult to find the series coefficients. Then a residual power series approach is brought to find the coefficients in sequential form as recurrence relation by using transformed functions. For this, a reliable, but not common, method is established, which can be used by differentiating the nth partial sum of that power series in $(n-1)$ times to find its nth ordered coefficients, which is the residual power series method. In general, for handling some fractional linear problems, the ordinary derivatives are updated to fractional-order derivatives. Laplace transform with residual power series approach is established for handling such fractional linear problems. Therefore, some of the important models arising in different branches of mathematics, physics, as well as engineering fields are solved by using Laplace transform with residual power series method analytically.

The aim of this work is to increase the reliability and accuracy of the RPSM by using the Laplace transform in the methodology of this problem. Consider a fractional-order differential equation of the following form:

$$D_t^\alpha u(x, t) + L[x] u(x, t) + \text{NL}[x] u(x, t) = 0 \qquad (9.1)$$

for $t > 0$, $x \in \mathcal{R}$, $0 < \alpha \leq 1$

with initial condition, $u(x, 0) = f_0(x)$ $\qquad (9.2)$

$$\text{and exact solution } u(x,t) = g(x,t) \tag{9.3}$$

where $D_t^\alpha = \frac{\partial^\alpha}{\partial t^\alpha}$, $L[x]$ is the linear operator in x, $NL[x]$ is the general nonlinear operator in x, and $u(x,t)$ is taken as continuous function.

This equation is a linear or nonlinear fractional-order differential equation and is solved by using Laplace transform with residual power series method. There are so many reliable, popular, as well as efficient numerical and analytic methods having more reliability to find the solutions of such problems. Out of them, residual power series method [17], homotopy perturbation and analysis methods [18], differential transform method [19], iterative method [20], Adomian decomposition method [21], different forms of fractional power series representation [22], and other methods have been used for solving fractional-order differential equations.

The manuscript is arranged as follows. In Section 9.1, the introduction is given. In Section 9.2, the preliminaries are mentioned. In Section 9.3, research methodology of the Laplace transform with residual power series method for solution of fractional-order one-dimensional differential equations is explained. In Section 9.4, the numerical solution of fractional-order one-dimensional differential equations by using Laplace transform with residual power series approach is given. Finally, the conclusion of the research topic is outlined in Section 9.5.

9.2 Preliminaries

Laplace transform:

Let $f(t)$ be a function defined for all $t>0$. Then the Laplace transform of $f(t)$ denoted by $\mathcal{L}\{f(t)\}$ is defined as $\mathcal{L}\{f(t)\} = \int_0^\infty e^{-pt} f(t)\, dt$ provided that the integral exists with parameter t.

It is clear that the Laplace transform is a function of p and is denoted by $\overline{f}(p)$.

$$\therefore \overline{f}(p) = \mathcal{L}\{f(t)\} = \int_0^\infty e^{-pt} f(t)\, dt.$$

Again, the function $f(t)$ is called the inverse Laplace transform of $\overline{f}(p)$. The Laplace transform exists clearly if the following two conditions are satisfied by the function $f(t)$:

i) $f(t)$ is continuous;
ii) $\lim_{t \to \infty} e^{-pt} f(t)$ is finite.

It is remarkable that conditions are sufficient but not necessary.

Laplace transforms of some standard functions:

i) $\mathcal{L}(1) = \frac{1}{p}$ for $p > 0$.

ii) $\mathcal{L}(t^n) = \frac{n!}{p^{n+1}} = \frac{\Gamma(n+1)}{p^{n+1}}$, where $n = 0, 1, 2 \ldots$

iii) $\mathcal{L}(e^{at}) = \frac{1}{p-a}$ (for $p > 0$).

iv) $\mathcal{L}(\sin at) = \frac{a}{p^2+a^2}$ ($p > 0$).

v) $\mathcal{L}(\cos at) = \frac{p}{p^2+a^2}$ ($p > 0$).

vi) $\mathcal{L}(\sinh at) = \frac{a}{p^2-a^2}$ ($p > |a|$).

vii) $\mathcal{L}(\cosh at) = \frac{p}{p^2-a^2}$ ($p > |a|$).

Properties of Laplace transform:

1) Linearity property: If $a, b,$ and c are constants and $f, g,$ and h are functions of t, then $\mathcal{L}\{af(t) + bg(t) - ch(t)\} = a\mathcal{L}\{f(t)\} + b\mathcal{L}\{g(t)\} - c\mathcal{L}\{h(t)\}$.

2) First shifting property: If $\mathcal{L}\{f(t)\} = \overline{f}(p)$, then $\mathcal{L}\{e^{at}f(t)\} = \overline{f}(p-a)$. With the help of this property, it can be determined that if $\overline{f}(p)$ is the Laplace transform of $f(t)$, then the Laplace transform of $e^{at}f(t)$ can be simply written by replacing p by $p - a$ to get $\overline{f}(p-a)$.

Again this property gives us the following useful formulae:

i) $\mathcal{L}(e^{at}) = \frac{1}{p-a}$ ($p > 0$)

ii) $\mathcal{L}(e^{at}t^n) = \frac{n!}{(p-a)^{n+1}} = \frac{\Gamma(n+1)}{(p-a)^{n+1}}$, where $n = 0, 1, 2, \ldots$

iii) $\mathcal{L}(e^{at}\sin bt) = \frac{b}{(p-a)^2+b^2}$

iv) $\mathcal{L}(e^{at}\cos bt) = \frac{p-a}{(p-a)^2+b^2}$

v) $\mathcal{L}(e^{at}\sinh bt) = \frac{b}{(p-a)^2-b^2}$

vi) $\mathcal{L}(e^{at}\cosh bt) = \frac{p-a}{(p-a)^2-b^2}$, where $p > a$ in each case.

Power series:

An infinite series of the form
$$\sum_{n=0}^{\infty} c_n(x - x_0)^n = c_0 + c_1(x - x_0) + c_2(x - x_0)^2 + \ldots$$ is known as a general power series in $x - x_0$. In particular, an infinite series $\sum_{n=0}^{\infty} c_n x^n = c_0 + c_1 x + c_2 x^2 + \ldots$ is known as power series in x.

Convergence of power series:

The power series $\sum_{n=0}^{\infty} c_n(x-x_0)^n$ converges (absolutely) for $|x| < R$, where $R = lim_{n \to \infty} \left|\frac{c_n}{c_{n+1}}\right|$, provided that the limit exists.

Radius of convergence and interval of convergence:

If a given power series does not converge everywhere or nowhere, then a definite number $R > 0$ exists such that the given power series converges (absolutely) for every $|x| < R$ and diverges for every $|x| > R$, and such a number R is known as radius of convergence.

The open interval $(-R, R)$ is known as an interval of convergence of the given power series.

9.3 Methodology

The method of solution of fractional-order logistic differential equation by using Laplace transform with residual power series method can be done in the following steps:

Step 1: Applying the Laplace transform on fractional-order differential equation (9.1), it becomes

$$\mathcal{L}\left[D_t^\alpha u(x, t) + L[x]u(x, t) + \text{NL}[x]u(x, t)\right] = 0. \quad (9.4)$$

From Laplace transform of fractional derivatives using the relation, $\mathcal{L}[D_t^\alpha u(x,t)] = s^\alpha \mathcal{L}[u(x,t)] - s^{\alpha-1}u(x,0)$ on eqn (9.4), then it can be framed as

$$U(x, s) = \frac{f_0(x)}{s} - L[x]U(x, s) - \text{NL}[x]\mathcal{L}[(\mathcal{L}^{-1}[U(x, s)])] \quad (9.5)$$

where $U(x, s) = \mathcal{L}[u(x, t)]$.

Step 2: The transformed function $U(x, s)$ can be written as

$$U(x, s) = \sum_{n=0}^{\infty} \frac{f_n(x)}{s^{n\alpha+1}}. \quad (9.6)$$

Also the kth - truncated series of relation (9.6) can be written as

$$U_k(x, s) = \sum_{n=0}^{k} \frac{f_n(x)}{s^{n\alpha+1}}$$

i.e., $U_k(x,s) = \dfrac{f_0(x)}{s} + \sum_{n=1}^{k} \dfrac{f_n(x)}{s^{n\alpha+1}}.$ (9.7)

Again the kth - Laplace residual function [23] is

$$\mathcal{L}\text{Res}_k(x,s) = U_k(x,s) - \dfrac{f_0(x)}{s} + L[x]U_k(x,s)$$
$$+ \text{NL}[x]\mathcal{L}[(\mathcal{L}^{-1}[U_k(x,s)])^r]. \quad (9.8)$$

Substituting the kth truncated series (9.7) into the kth Laplace residual function (9.8), it becomes

$$\mathcal{L}\text{Res}_k(x,s) = \dfrac{f_0(x)}{s} + \sum_{n=1}^{k} \dfrac{f_n(x)}{s^{n\alpha+1}} - \dfrac{f_0(x)}{s} + L[x]\{\dfrac{f_0(x)}{s} + \sum_{n=1}^{k} \dfrac{f_n(x)}{s^{n\alpha+1}}\}$$
$$+ \text{NL}[x]\mathcal{L}[\left(\mathcal{L}^{-1}\left[\dfrac{f_0(x)}{s} + \sum_{n=1}^{k} \dfrac{f_n(x)}{s^{n\alpha+1}}\right]\right)^2]$$

$$\mathcal{L}\text{Res}_k(x,s) = \sum_{n=1}^{k} \dfrac{f_n(x)}{s^{n\alpha+1}} + L[x]\{\dfrac{f_0(x)}{s} + \sum_{n=1}^{k} \dfrac{f_n(x)}{s^{n\alpha+1}}\}$$
$$+ \text{NL}[x]\mathcal{L}\left[\left(\mathcal{L}^{-1}\left[\dfrac{f_0(x)}{s} + \sum_{n=1}^{k} \dfrac{f_n(x)}{s^{n\alpha+1}}\right]\right)^2\right]. \quad (9.9)$$

Step 3: By solving the following relation recursively [24], the coefficients $f_n(x)$ can be obtained:

$$\lim_{s\to\infty} s^{k\alpha+1}\mathcal{L}\text{Res}_k(x,s) = 0 \text{ for } 0 < \alpha \leq 1, k = 1, 2, 3, \ldots. \quad (9.10)$$

Following are some useful relations that are used in standard RPSM [25]:
i) $\mathcal{L}\text{Res}(x,s) = 0$ and $\lim_{k\to\infty} \mathcal{L}\text{Res}_k(x,s) = \mathcal{L}\text{Res}(x,s),$ for $s > 0$.
ii) $\lim_{s\to\infty} s\mathcal{L}\text{Res}(x,s) = 0$ gives $\lim_{s\to\infty} s\mathcal{L}\text{Res}_k(x,s) = 0$.
iii) $\lim_{s\to\infty} s^{k\alpha+1}\mathcal{L}\text{Res}(x,s) = \lim_{s\to\infty} s^{k\alpha+1}\mathcal{L}\text{Res}_k(x,s) = 0$ for $0 < \alpha \leq 1$.

Step 4: Finally, applying the inverse Laplace transform to $U_k(x,s)$ for obtaining the kth approximate solution $u_k(x,t)$ of fractional-order differential equations is obtained.

This is one of the reliable methods for the solution of fractional differential equations [26].

9.4 Numerical Solutions

Consider the linear Schrödinger equation

$$D_t^\alpha u + iu_{xx} = 0 \tag{9.11}$$

with initial condition

$$u(x,0) = 1 + \cosh(2x) \tag{9.12}$$

and exact solution is $u(x,t) = 1 + E_\alpha(-4it^\alpha)\cosh(2x)$ where $E_\alpha(z) = \sum_{k=0}^{\infty} \frac{z^k}{\Gamma(1+k\alpha)}$, $z = -4it^\alpha$

$$\text{and } u(x,t) = 1 + \cosh(2x)\,e^{-4it} \text{ for } \alpha = 1 \text{ with } i^2 = -1. \tag{9.13}$$

Taking Laplace transform on eqn (9.11), we get

$$\mathcal{L}[D_t^\alpha u] + \mathcal{L}[iu_{xx}] = 0. \tag{9.14}$$

From Laplace transform of fractional derivatives using the relation $\mathcal{L}[D_t^\alpha u(x,t)] = s^\alpha \mathcal{L}[u(x,t)] - s^{\alpha-1} u(x,0)$ in eqn (9.14), then it can be framed as

$$s^\alpha \mathcal{L}[u(x,t)] - s^{\alpha-1} u(x,0) + \mathcal{L}[iu_{xx}] = 0$$

$$s^\alpha \mathcal{L}[u(x,t)] = s^{\alpha-1} u(x,0) - i\mathcal{L}[u_{xx}]$$

$$U(x,s) = \frac{f_0(x)}{s} - i\frac{1}{s^\alpha}\{U(x,s)\}_{xx} \tag{9.15}$$

where $U(x,s) = \mathcal{L}[u(x,t)]$ and $u(x,0) = f_0(x)$.
The transformed function $U(x,s)$ can be written as

$$U(x,s) = \sum_{n=0}^{\infty} \frac{f_n(x)}{s^{n\alpha+1}}. \tag{9.16}$$

Also the kth truncated series of relation (9.16) can be written as

$$U_k(x,s) = \sum_{n=0}^{k} \frac{f_n(x)}{s^{n\alpha+1}}$$

i.e., $$U_k(x,s) = \frac{f_0(x)}{s} + \sum_{n=1}^{k} \frac{f_n(x)}{s^{n\alpha+1}}. \tag{9.17}$$

Again the kth Laplace residual function of eqn (9.15) is

$$\mathcal{L}\text{Res}_k(x,s) = U_k(x,s) - \frac{f_0(x)}{s} - i\frac{1}{s^\alpha}\{U_k(x,s)\}_{xx}. \qquad (9.18)$$

To find the values of $f_k(x)$, $k = 1, 2, 3, \ldots$, substitute the kth truncated series (9.17) into the kth Laplace residual function (9.18), and we get

$$\mathcal{L}\text{Res}_k(x,s) = \frac{f_0(x)}{s} + \sum_{n=1}^{k}\frac{f_n(x)}{s^{n\alpha+1}} - \frac{f_0(x)}{s} - i\frac{1}{s^\alpha}\{\frac{f_0(x)}{s} + \sum_{n=1}^{k}\frac{f_n(x)}{s^{n\alpha+1}}\}_{xx}$$

$$= \sum_{n=1}^{k}\frac{f_n(x)}{s^{n\alpha+1}} - i\frac{1}{s^\alpha}\{\frac{f_0(x)}{s} + \sum_{n=1}^{k}\frac{f_n(x)}{s^{n\alpha+1}}\}_{xx}. \qquad (9.19)$$

For $k = 1$ from eqn (9.19), the first Laplace residual function is

$$\mathcal{L}\text{Res}_1(x,s) = \frac{f_1(x)}{s^{\alpha+1}} - i\frac{1}{s^\alpha}\{\frac{1+\cosh(2x)}{s} + \frac{f_1(x)}{s^{\alpha+1}}\}_{xx}$$

$$[\because u(x,0) = 1 + \cosh(2x) = f_0(x)]$$

$$= \frac{f_1(x)}{s^{\alpha+1}} - \frac{4i\cosh(2x)}{s^{\alpha+1}} - i\frac{(f_1(x))_{xx}}{s^{2\alpha+1}}.$$

Now, the relation $\lim_{s \to \infty}(s^{\alpha+1}\mathcal{L}\text{Res}_1(x,s)) = 0$ for $k = 1$ gives

$$f_1(x) - 4i\cosh(2x) = 0, \text{ i.e., } f_1(x) = 4i\cosh(2x).$$

For $k = 2$, from eqn (9.19), the second Laplace residual function is

$$\mathcal{L}\text{Res}_2(x,s) = \frac{f_1(x)}{s^{\alpha+1}} + \frac{f_2(x)}{s^{2\alpha+1}} - i\frac{1}{s^\alpha}\{\frac{1+\cosh(2x)}{s} + \frac{f_1(x)}{s^{\alpha+1}} + \frac{f_2(x)}{s^{2\alpha+1}}\}_{xx}$$

$$= \frac{4i\cosh(2x)}{s^{\alpha+1}} + \frac{f_2(x)}{s^{2\alpha+1}} - i\frac{1}{s^\alpha}\{\frac{1+\cosh(2x)}{s} + \frac{4i\cosh(2x)}{s^{\alpha+1}} + \frac{f_2(x)}{s^{2\alpha+1}}\}_{xx}$$

$$= \frac{4i\cosh(2x)}{s^{\alpha+1}} + \frac{f_2(x)}{s^{2\alpha+1}} - \frac{4i\cosh(2x)}{s^{\alpha+1}} + \frac{16\cosh(2x)}{s^{2\alpha+1}} - \frac{i(f_2(x))_{xx}}{s^{3\alpha+1}}$$

$$= \frac{f_2(x)}{s^{2\alpha+1}} + \frac{16\cosh(2x)}{s^{2\alpha+1}} - \frac{i(f_2(x))_{xx}}{s^{3\alpha+1}}.$$

Now the relation $\lim_{s \to \infty}(s^{2\alpha+1}\mathcal{L}\text{Res}_2(x,s)) = 0$ for $k = 2$ gives us that

$$f_2(x) + 16\cosh(2x) = 0, \text{ i.e., } f_2(x) = -16\cosh(2x).$$

9.4 Numerical Solutions

For $k = 3$, from eqn (9.19), the third Laplace residual function is

$$\mathcal{L}\text{Res}_3(x, s) = \frac{f_1(x)}{s^{\alpha+1}} + \frac{f_2(x)}{s^{2\alpha+1}} + \frac{f_3(x)}{s^{3\alpha+1}} - i\frac{1}{s^\alpha}\{\frac{1+\cosh(2x)}{s}$$

$$+ \frac{f_1(x)}{s^{\alpha+1}} + \frac{f_2(x)}{s^{2\alpha+1}} + \frac{f_3(x)}{s^{3\alpha+1}}\}_{xx}$$

$$= \frac{4i\cosh(2x)}{s^{\alpha+1}} - \frac{16\cosh(2x)}{s^{2\alpha+1}} + \frac{f_3(x)}{s^{3\alpha+1}} - i\frac{1}{s^\alpha}\{\frac{1+\cosh(2x)}{s} + \frac{4i\cosh(2x)}{s^{\alpha+1}}$$

$$- \frac{16\cosh(2x)}{s^{2\alpha+1}} + \frac{f_3(x)}{s^{3\alpha+1}}\}_{xx}$$

$$= \frac{4i\cosh(2x)}{s^{\alpha+1}} - \frac{16\cosh(2x)}{s^{2\alpha+1}} + \frac{f_3(x)}{s^{3\alpha+1}} - i\{\frac{1}{s^\alpha}\frac{4\cosh(2x)}{s} + \frac{16i\cosh(2x)}{s^{\alpha+1}}$$

$$- \frac{64\cosh(2x)}{s^{2\alpha+1}} + \frac{(f_3(x))_{xx}}{s^{3\alpha+1}}\}$$

$$= \frac{4i\cosh(2x)}{s^{\alpha+1}} - \frac{16\cosh(2x)}{s^{2\alpha+1}} + \frac{f_3(x)}{s^{3\alpha+1}} - \frac{4i\cosh(2x)}{s^{\alpha+1}} + \frac{16\cosh(2x)}{s^{2\alpha+1}}$$

$$+ \frac{64i\cosh(2x)}{s^{3\alpha+1}} - \frac{i(f_3(x))_{xx}}{s^{4\alpha+1}}$$

$$= \frac{f_3(x)}{s^{3\alpha+1}} + \frac{64i\cosh(2x)}{s^{3\alpha+1}} - \frac{i(f_3(x))_{xx}}{s^{4\alpha+1}}.$$

Now the relation $\lim_{s\to\infty}(s^{3\alpha+1}\mathcal{L}\text{Res}_3(x, s)) = 0$ for $k = 3$ gives us that

$$f_3(x) + 64i\cosh(2x) = 0, \text{i.e., } f_3(x) = -64i\cosh(2x).$$

Hence, the Laplace residual power series solution of a given equation in infinite form is

$$U(x, s) = \frac{1+\cosh(2x)}{s} + 4i\cosh(2x)\frac{1}{s^{\alpha+1}} - 16\cosh(2x)\frac{1}{s^{2\alpha+1}}$$

$$- 64i\cosh(2x)\frac{1}{s^{3\alpha+1}} - \cdots. \qquad (9.20)$$

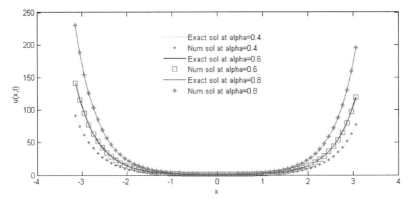

Figure 9.1 Comparison of solution of Example 1 for different values of α.

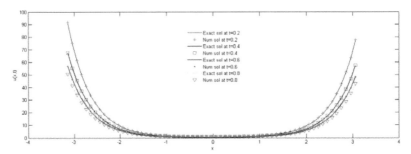

Figure 9.2 Comparison of solution of Example 1 for different time levels at $\alpha = 0.4$.

Finally, by taking inverse Laplace transform in eqn (9.20), we get the required solution of a given equation by using the Laplace transform with RPSM as

$$u(x,t) = 1 + \cosh(2x) + 4i\cosh(2x)\frac{t^{\alpha}}{\alpha!} - 16\cosh(2x)\frac{t^{2\alpha}}{(2\alpha)!}$$
$$- 64i\cosh(2x)\frac{t^{3\alpha}}{(3\alpha)!} - \ldots \qquad (9.21)$$

The solution of the example has been calculated by taking the values of α as 0.4, 0.6, and 0.8 for domain $[-\pi, \pi]$ for $t = 0.2$, and the solution is presented in Figure 9.1. The solution is depending on the number of terms in the power series solution. The graph is plotted by taking 10 terms and the results are satisfactory and are comparable with the exact solutions. In Figure 9.2, the solution is presented for the different values of t as 0.2, 0.4, 0.6, and 0.8 at

$\alpha = 0.4$. It can be seen that the solution obtained by the method is satisfying the exact solution.

9.5 Conclusion

In this chapter, a new and reliable scheme is constructed for the solution of one-dimensional fractional-order differential equations by using Laplace transform with RPSM. The advantage of the present method is to decrease the computational work required for finding the solution of one-dimensional fractional-order differential equations by using Laplace transform in residual power series form such that the coefficients of which are determined in the above successive steps. The one-dimensional fractional-order differential equations are solved by using the Laplace transform with residual power series approaches; it proved its ability to solve linear and nonlinear fractional-order differential equations with sufficient accuracy and reliable calculation steps.

References

[1] Yu Zhang, Amit Kumar, Sunil Kumar, Dumitru Baleanu, Xiao-Jun Yang, "Residual power series method for time-fractional Schríodinger equations", Journal of Nonlinear Science and Applications, 2016 (5821-5829).

[2] A. A. Kilbas, H. M. Srivastava, J. J. Trujillo, "Theory and applications of fractional differential equations", North Holland Mathematics Studies, Elsevier Science B. V., Amsterdam, 2006 (1-2.2).

[3] I. Podlubny, "Fractional differential equations, an introduction to fractional derivatives, fractional differential equations, to methods of their solution and some of their applications", Mathematics in Science and Engineering, Academic Press, Inc., San Diego, CA, 1999 (1- 2.2).

[4] X.-J. Yang, D. Baleanu, H. M. Srivastava, "Local fractional integral transforms and their applications", Elsevier/Academic Press, Amsterdam, 2016 (1-2.2).

[5] S. Abbasbandy, "The application of homotopy analysis method to nonlinear equations arising in heat transfer", Phys. Lett. A, 2006 (109–113).

[6] S. Kumar, "A numerical study for the solution of time fractional nonlinear shallow water equation in oceans", Z. Naturforschung A, 2013 (547–553).

[7] S. Kumar, "A new analytical modelling for fractional telegraph equation via Laplace transform", Appl. Math. Model., 2014 (3154–3163).
[8] S. Kumar, M. M. Rashidi, "New analytical method for gas dynamic equation arising in shock fronts", Comput. Phys. Commun., 2014 (1947–1954)
[9] S. Momani, Z. Odibat, "Analytical solution of a time-fractional Navier-Stokes equation by Adomian decomposition method", Appl. Math. Comput., 2006 (488–494).
[10] Z.M.Odibat, S.Momani, "Application of variational iteration method to nonlinear differential equations of fractional order", Int. J. Nonlinear Sci. Number Simul., 2006 (27-34).
[11] X.-J.Yang, D. Baleanu, "Fractal heat conduction problem solved by local fractional variation iteration method", Therm. Sci., 2013 (625–628).
[12] X.-J.Yang, D.Baleanu, M.P.Lazarevic, M.S.Cajic, "Fractal boundary value problems for integral and differential equations with local fractional operators", Therm. Sci., 2013 (959–966).
[13] X.-J.Yang, D.Baleanu, W.-P.Zhong, "Approximate solutions for diffusion equations on Cantor space-time", Proc. Rom. Acad. Ser. A Math. Phys. Tech. Sci. Inf. Sci., 2013 (127–133).
[14] X.J.Yang, H. M.Srivastava, C.Cattani, "Local fractional homotopy perturbation method for solving fractal partial differential equations arising in mathematical physics", Rom. Rep. Phys., 2015, (752–761).
[15] X.J.Yang, J.A.Tenreiro Machado, D. Baleanu, C. Cattani, "On exact travelling-wave solutions for local fractional Korteweg-de Vries equation", Chaos, 2016 (205-210).
[16] Alquran M., Ali M. Alsukhour, M. Jaradat, I., "Promoted residual power series technique with Laplace transform to solve some time-fractional problems arising in physics", Results in Physics, 2020 (103667).
[17] Alquran, M., Jaradat, I., "Delay- asymptotic solutions for the time-fractional delay-type wave equation", Physica A, 2019 (121275).
[18] Ganjiani, M., "Solution of non-linear fractional differential equations using homotopy analysis method", Appl Math Comput, 2010 (634-4).
[19] Yousef, F., Alquran, M., Jaradat, I., Momani, S., Baleanu, D., "Ternary-fractional differential transform schema: theory and application", Adv Difference Equ, 2019 (2019:197).
[20] Odibat, Z., Momani, S., "Application of variational iteration method to non-linear differential equations of fractional order", Int J Nonlinear Sci Number Simul, 2006 (27-34).

[21] Ray, SS., Bera, RK., "Analytical solution of a fractional diffusion equation by Adomian decomposition method", Appl Math Comput, 2006 (329-36).
[22] Jaradat, I., Al-Dolat, M., Al-Zoubi, K., Alquran, M., "Theory and applications of a more general form for fractional power series expansion", Chaos Solitons Fractals 2018 (107-10).
[23] Eriqat T., El-Ajou A., Moa'ath, No., Al-Zhour Z., Momani,S., "A new attractive analytic approach for solution of linear and non-linear Neutral Fractional Pantograph equations", Choas Solitons Fractals, 2020 (109957).
[24] Komashynska I., Al-Smadi M., Abu Arqub O., Momani S., "An efficient analytical method for solving singular initial value problems of non-linear systems", Appl Math Inf Sci 2016 (647-56).
[25] El-Ajou A., Abu Arqub O., Al-Smadi M., "A general form of the generalised Taylor's formula with some applications", Appl.Math.Comput, 2015 (256:851-9).
[26] Eriqat T., El-Ajou A., Moa'ath N.O., Al-Zhour Z., Momani S, "A new attractive analytic approach for solutions of linear and nonlinear Neutral fractional Pantograph equations", Chaos Solitons Fractals, 2020 (109957).

Index

2D nonlinear time-fractional coupled Burgers' equation 103, 114

A
astrophysics equations 45
adomian decomposition method 103, 137

B
bessel function 48
B-spline 21, 56, 64, 67

C
convection–diffusion equation 21, 80
cubic spline 63
crystal lattice 117, 121
computational simulations 118

D
differential quadrature method 21, 31, 63, 85

E
electric field 1, 15, 119

H
hypergeometric functions 45

K
Khater II method 1, 6, 121
Klein–Gordon equation 85

L
legendre polynomials 46
Laplace transforms 136
Laplace residual function 136, 140, 142

M
modified cubic B-splines 85

N
new hybrid cubic b-spline 21
numerical techniques 22, 37, 64, 79

P
polytropic fluid 45, 47
plasma physics 117, 120

Q
quintic spline 63
quadratic spline 63
quartic spline 63

R
Rangwala–Rao equation 1, 15
ranking function 37, 39, 41
residual power series method 135, 137

S
solitary wave 1, 6, 122
SSP-RK43 formulae 23

symmetric trapezoidal number 37
spline 23, 56, 63, 87
sixth-order compact finite difference scheme 85
Sumudu transform 103, 104, 105

Sumudu ADM 103
soliton wave 1, 15, 122
stability 118, 120
Schrödinger differential equation 136

About the Editors

Dr. Geeta Arora received her the Ph.D. degree in Mathematics from IIT Roorkee, India in 2011. She is currently serving as a Professor in the Department of Mathematics at Lovely Professional University, Punjab, India. She has more than 11 years of research and teaching experience and has taught several core courses in applied mathematics at undergraduate, postgraduate, and doctorate levels. She has published around 50 research papers in both international and national Journals. She has authored eight book chapters in national and international publications. Also, she has written a book on Quick calculations Calculations in Mathematics and another on Statistics. Her area of research is development of numerical methods and statistics. She has received the Research Appreciation Award in 2017 and 2019 by Lovely Professional University, Punjab, India. She has conducted several workshops on MATLAB and Vedic mathematics for students and faculty members within and outside the university campus. She has supervised five students and is currently guiding five Ph.D. scholars.

Prof. Mangey Ram received the Ph.D. degree major in Mathematics and minor in Computer Science from G. B. Pant University of Agriculture and Technology, Pantnagar, India in 2008. He has been a faculty member for around thirteen years and has taught several core courses in pure and applied mathematics at undergraduate, postgraduate, and doctorate levels. He is currently the a Research Professor at Graphic Era (Deemed to be University), Dehradun, India and visiting professor at Peter the Great St. Petersburg Polytechnic University, Saint Petersburg, Russia. Before joining the Graphic Era, he was a Deputy Manager (Probationary Officer) with Syndicate Bank for a short period. He is the Editor-in-Chief of *International Journal of Mathematical, Engineering and Management Sciences*; *Journal of Reliability and Statistical Studies*; *Journal of Graphic Era University*; Series Editor of six book series with Elsevier, CRC Press-A Taylor and Frances Group, Walter De Gruyter Publisher Germany, and River Publisher; and the Guest Editor and Associate Editor for various journals. He has published more than 300 plus publications (journal articles/books/book chapters/conference articles) in IEEE, Taylor & Francis, Springer Nature, Elsevier, Emerald, World Scientific, and many other national and international journals and conferences. Also, he has published more than 60 books (authored/edited) with international publishers like Elsevier, Springer Nature, CRC Press-A Taylor and Frances Group, Walter De Gruyter Publisher Germany, and River Publisher. His fields of research are reliability theory and applied mathematics. Dr. Ram is a Senior Member of the IEEE, and Senior Life Member of Operational Research Society of India, Society for Reliability Engineering, Quality and Operations Management in India, Indian Society of Industrial and Applied Mathematics. He has been a member of the organizing committee of a number of international and national conferences, seminars, and workshops. He has been conferred with *"Young Scientist Award"* by the Uttarakhand State Council for Science and Technology, Dehradun, in 2009. He has been awarded the *"Best Faculty Award"* in 2011; "Research Excellence Award" in 2015; and *"Outstanding Researcher Award"* in 2018 for his significant contribution in academics and research at Graphic Era Deemed to be University, Dehradun, India. Recently, he has been received the ""*Excellence in Research of the Year-2021 Award*" by the Honourable Chief Minister of Uttarakhand State, India, and "Emerging Mathematician of Uttarakhand" state award by the Director, Uttarakhand Higher Education.